Letts

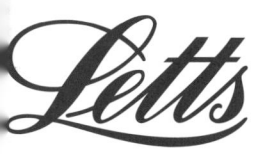

CAMBRIDGE IGCSE® MATHS

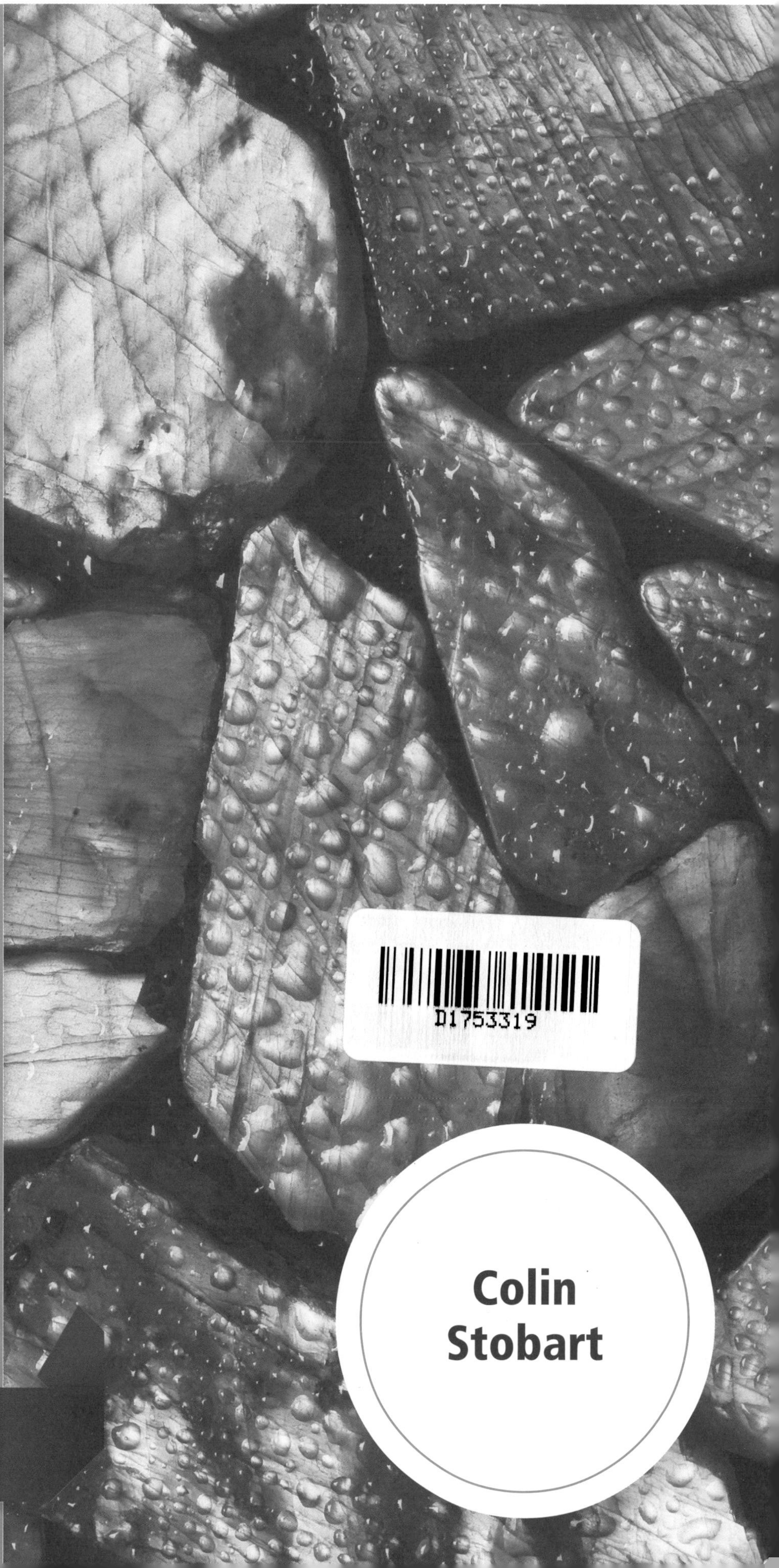

Colin Stobart

Revision Guide

ACKNOWLEDGEMENTS

Cover photo and page 1: © J. Palys / Shutterstock

Published by Letts Educational
An imprint of HarperCollins*Publishers*
The News Building
1 London Bridge Street
London
SE1 9GF

ISBN 978-0-00-821034-2

First published 2018

10 9 8 7 6 5 4 3 2 1

© HarperCollins*Publishers* Limited 2018

®IGCSE is a registered trademark.
All exam-style questions, related example answers, marks awarded and comments that appear in this book were written by the author. In examinations, the way marks would be awarded to questions and answers like these may be different.

Colin Stobart asserts his moral right to be identified as the author of this work.

All rights reserved. No part of this publication may be reproduced, stored in a retrieval system, or transmitted in any form or by any means, electronic, photocopying, recording or otherwise, without the prior written permission of Letts Educational.

British Library Cataloguing in Publication Data
A CIP record for this book is available from the British Library.

Commissioned by Gillian Bowman
Project managed by Rachel Allegro
Copyedited by Joan Miller
Proofread by Anna Clark and Paul Winters
Cover design by Paul Oates
Typesetting by QBS
Production by Natalia Rebow and Lyndsey Rogers
Printed and bound in China by RR Donnelley APS

MIX
Paper from responsible sources
FSC™ C007454

This book is produced from independently certified FSC paper to ensure responsible forest management.

For more information visit: www.harpercollins.co.uk/green

Contents

Introduction	p.5
Number	p.6
Number	p.6
Fractions and percentages 1	p.10
Fractions and percentages 2	p.13
The four rules	p.16
Directed numbers	p.18
Squares and cubes	p.19
Ordering and set notation	p.21
Ratio, proportion and rate 1	p.23
Ratio, proportion and rate 2	p.25
Estimation and limits of accuracy 1	p.27
Estimation and limits of accuracy 2	p.29
Standard form	p.31
Applying number and using calculators	p.33
Exam-style practice questions	p.37
Algebra	p.41
Algebraic representation and formulae	p.41
Algebraic manipulation 1	p.44
Algebraic manipulation 2	p.47
Solutions of equations and inequalities 1	p.49
Simultaneous equations	p.52
Solutions of equations and inequalities 2	p.54
Graphs in practical situations	p.56
Straight-line graphs	p.59
Graphs of functions	p.63
Number sequences	p.68
Indices	p.72
Variation	p.74
Linear programming	p.76
Functions	p.79
Exam-style practice questions	p.81

Geometry — p.87
- Angle properties — p.87
- Geometrical terms and relationships — p.92
- Geometrical constructions — p.98
- Trigonometry — p.100
- Mensuration — p.108
- Symmetry — p.114
- Vectors — p.116
- Transformations — p.118
- Exam-style practice questions — p.123

Statistics — p.130
- Statistical representation — p.130
- Statistical measures — p.136
- Probability — p.143
- Exam-style practice questions — p.148

Answers — p.152
Glossary — p.162

Introduction

This revision guide is intended to support your studies for *Cambridge IGCSE Mathematics*. Its content is linked to the syllabus, providing revision tips and plenty of examples and revision questions, all designed to help you to prepare for the examinations.

However, it is important to remember that this is a revision guide. Its structure closely follows the *Collins Cambridge IGCSE Mathematics Student Book* and covers all of the syllabus content, but it does not go into the depth and detail you might get from your own notes or from the textbook. It is therefore a good idea to use this book as one of the tools to prepare you for the examinations, but it should not be the only tool in your toolkit.

The book is split into four units: Number, Algebra, Geometry, and Statistics. Each unit contains sections that fully cover the syllabus content for that unit. At the end of each section within the unit there is a **Quick test**, covering the topics in that section, to help you check that you have understood the main aspects of the topic area. At the end of each unit there is a set of **Examination-style practice questions** that cover every aspect of the unit's syllabus content. The answers to all the Quick test and Examination-style practice questions are at the back of the book. We have included a marking-scheme with the Exam-style practice questions to guide you. Where questions are worth 1 mark, it is acceptable for you to just provide the answer as that is what the mark is awarded for. When questions are allocated more than 1 mark, you will need to supply more than just the answer. Marks are awarded for 'method' and showing how you arrive at the answer. This might mean: substituting values into an expression or equation; identifying values in the question and creating a workable equation; correctly simplifying an expression or equation in order to move to the next stage of the solution. Marks are also awarded for accuracy – these marks are given, for example, if an answer is required to a specified number of decimal places or significant figures. Be careful of questions such as, 'Suri says that the plants would be cheaper at Gardeners Paradise. Is she correct?' With these questions, marks are given for your answer of yes/no, but also for supporting evidence. Generally, you need to ensure that you have an identifiable step/reason for each mark that is available.

Throughout the units you will find some sections and questions have a shaded blue background. These sections are specific to the Extended syllabus; if you are working towards the Core examinations you will not have studied these areas as part of your course.

At the back of the book is a glossary, making it easy to find the meanings of the key words (in bold) that you need to know.

The **Revision tips** that appear in each unit offer additional advice on methods to use and facts that you need to remember. They might also highlight particular concepts that typically confuse students.

Finally, think about how you revise. While you may want to focus on your own learning and work through past examination papers or questions, it is also a good idea to meet up with other students preparing for the examinations and, for at least some of the time, revise together. You can help each other with aspects of the course, and devise quizzes to test each other. Working with friends can prove very useful for revision since we all have tips, shortcuts and memory aids that help us in our own work. Most importantly, revision with someone else is usually more fun and productive than when you revise on your own.

If you work your way carefully through the materials presented in this book in combination with some of the other revision strategies mentioned above, you will be well prepared.

Good luck!

Number

Square numbers and roots

When you multiply a number (n) by itself ($n \times n$) you get the square of the number (n^2). The sequence of **square numbers** begins 1, 4, 9, 16, 25… ($1^2, 2^2, 3^2, 4^2, 5^2$…)

The square root of a number (n) is the number of which the square is n. It is written as $\sqrt{n}$. For example, the **square root** of 144 is written as $\sqrt{144}$, which is ±12.

Check: 12×12, or $12^2 = 144$, or -12×-12, or $-12^2 = 144$

Cube numbers and roots

When you multiply a number (n) by itself twice ($n \times n \times n$) you get the cube of the number (n^3). This sequence of **cube numbers** begins 1, 8, 27, 64, 125… ($1^3, 2^3, 3^3, 4^3, 5^3$…).

The cube root of a number (n) is the number of which the cube is n. It is written as $\sqrt[3]{n}$. For example, the **cube root** of 343 is written as $\sqrt[3]{343}$, which is 7.

Check: $7 \times 7 \times 7$, or $7^3 = 343$

Multiples

A **multiple** is the result of multiplying two numbers together. For example, $21 = 3 \times 7$, so 21 is a multiple of 3 and of 7.

The first five multiples of 3 are: 3, 6, 9, 12, 15

Check: 1×3, 2×3, 3×3, 4×3, 5×3

Factors

A **factor** is a whole number that will divide exactly into another number.

The factors of 24 are: 1, 2, 3, 4, 6, 8, 12, 24

The factors of 16 are: 1, 2, 4, 8, 16

The list of factors of a number always include 1 and the number itself.

> **Revision tip**
>
> Factors appear in the lists in pairs: 1 and 24, 2 and 12, 3 and 8, 4 and 6 – unless the number is a square number such as 16 where the factor 4 is only written once.

Prime numbers

Prime numbers only have two factors, 1 and the number itself. The number 1 is **not** prime because it only has one factor. The number 2 is the only even number that is prime.

The prime numbers between 10 and 40 are: 11, 13, 17, 19, 23, 29, 31, 37

Factorisation

Any whole number that is not prime can be written as the product of its **prime factors**, using a method called prime factorisation.

A simple way to find the prime factors is to create a prime-factor tree. For example, to find the prime factors of 630: $5 \times 126 = 630$, and $6 \times 21 = 126$ and so on… the factor circled at the end of each 'branch' is a prime number.

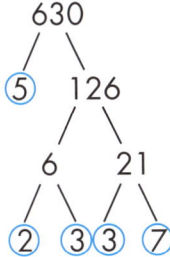

$$630 = 2 \times 3 \times 3 \times 5 \times 7$$
$$= 2 \times 3^2 \times 5 \times 7$$

An alternative method of finding the prime factors is to use the 'ladder' method. Keep dividing by prime numbers (start with 2, 3, 5…) to create new steps on the ladder. The prime numbers are in the left column and the final number is at the bottom.

```
2)630
3)315
3)105
5) 35
    7
```

so, $630 = 2 \times 3 \times 3 \times 5 \times 7$ as before.

Lowest common multiple (LCM)

The **lowest common multiple** (LCM) of two or more numbers is the smallest number into which the given numbers will all divide exactly.

One method of finding the LCM is to write a list of multiples for each of the given numbers. The LCM is the smallest number to appear in all of the lists.

A second method is to use the prime factorisation of each number. Lay out as in the following example and find the product of the factors in every column.

Example

Find the LCM of 45 and 54.

Method 1: 45: 45, 90, 135, 180, 225, 270, 315, …
54: 54, 108, 162, 216, 270, 324

Method 2: $45 = 3 \times 3 \times 5$
$54 = 2 \times 3 \times 3 \times 3$
LCM $= 2 \times 3 \times 3 \times 3 \times 5 = 270$

Number 7

Highest common factor (HCF)

The **highest common factor** (HCF) of two or more numbers is the biggest number that divides exactly into the given numbers.

One method of finding the HCF is to write out lists of factors of the given numbers and look for the biggest factor that appears in all the lists.

A second method is to use the list of numbers in the prime factorisations of the given numbers. Lay out as in the following example and find the product of the factors that appear in both rows.

> **Revision tip**
>
> Using prime factorisation speeds up the process of finding HCFs and LCMs.

> **Example**
> Find the HCF of 36 and 48.
> Method 1:   36: 1, 2, 3, 4, 6, 9, 12, 18, 36
>   48: 1, 2, 3, 4, 6, 8, 12, 16, 24, 48
> Method 2: $36 = 3 \times 3 \times 2 \times 2$
> $48 = 3 \times 2 \times 2 \times 2 \times 2 \times 2$
>   HCF $= 3 \times 2 \times 2 = 12$

Natural, integer, real, rational and irrational numbers

The set of positive whole numbers are the **natural numbers**: 1, 2, 3, 4, 5, …

Add the negative numbers to this set and you have the set of **integers**: …, −2, −1, 0, 1, 2, …

Add decimals to the set of integers and you have the set of **real numbers**, which may be of two kinds:

- **rational numbers**, which are integers, fractions or decimals, such as −3, $1\frac{2}{5}$, 2.35 or $\frac{23}{4}$.
- **irrational numbers**, which are those that cannot be written as a fraction, such as $\sqrt{15}$, $\sqrt{2}$ or π.

The reciprocal

The reciprocal of any number is 1 divided by the number.

For example:
- the reciprocal of 4 is 1 ÷ 4, which is $\frac{1}{4}$ = 0.25
- the reciprocal of 0.125 is 1 ÷ 0.125, which is 8

To find the reciprocal of a fraction, you invert it – swap around the numerator and denominator. A mixed number needs to be converted to an improper fraction before finding its reciprocal.

For example:
- the reciprocal of $\frac{1}{2}$ is $\frac{2}{1}$ = 2
- the reciprocal of $\frac{5}{4}$ is $\frac{4}{5}$
- The reciprocal of $3\frac{2}{5} = \frac{17}{5}$ is $\frac{5}{17}$

Remember to convert a mixed number to an improper fraction before finding its reciprocal.

Quick test

1. Here is a list of numbers: 1, 3, 6, 8, 9, 10, 12, 15
   a) Which two are square numbers?
   b) Which two are factors of 32?
   c) Which one is a prime number?
2. Find the highest common factor of: 16, 40 and 56.
3. Write 40 and 56 as the product of their prime factors and then use these to find the lowest common multiple of the two numbers.
4. Asako says all irrational numbers are those expressed with a $\sqrt{}$ (square root) sign. Give an example to show that this is not true.
5. Find the reciprocals of each of these:
   a) 16
   b) 0.4
   c) $\frac{5}{8}$
   d) $\frac{10}{3}$
   e) $3\frac{2}{3}$

Number

Fractions and percentages 1

Equivalent fractions, improper fractions and mixed numbers

Equivalent fractions represent the same quantity but are made up of different numerators (the number on the top) and denominators (the number on the bottom).

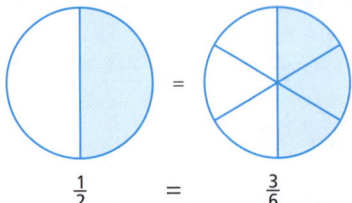

$$\frac{1}{2} = \frac{3}{6}$$

In this example $\frac{1}{2} = \frac{3}{6}$, so they are equivalent fractions.

Starting with $\frac{1}{2}$, both the numerator and denominator have been multiplied by 3, or starting with $\frac{3}{6}$ both the numerator and denominator have been divided by 3.

$\frac{1}{2}$ is $\frac{3}{6}$ written in its simplest form (or lowest terms). The only **common factor** of the numerator and denominator is 1.

Both $\frac{1}{2}$ and $\frac{3}{6}$ are proper fractions – the numerator is smaller than the denominator. A fraction, such as $\frac{5}{3}$, where the numerator is greater than the denominator is called an **improper fraction**.

A whole number and a fraction combined, such as $3\frac{4}{5}$, is a **mixed number**.

> **Revision tip**
>
> You may see the term *vulgar fraction*. This is just another term used to refer to a common fraction, one integer divided by another integer. Note that the denominator must not be 1 or 0.

Converting: decimals → fractions

Example

Convert 0.45 to a fraction.

Remember place value…

1	.	$\frac{1}{10}$	$\frac{1}{100}$
0	.	4	5

…and simplify: $\frac{45}{100} \rightarrow \frac{9}{20}$

Using the method in the above example, you could rewrite 0.645 as:

$\frac{6}{10} + \frac{4}{100} + \frac{5}{1000} = \frac{600}{1000} + \frac{40}{1000} + \frac{5}{1000} = \frac{645}{1000}$

Fractions written like this, using the normal decimal place-value system, are called decimal fractions.

Converting: fractions → decimals

For example, $\frac{5}{8}$, is a division calculation, $5 \div 8 \rightarrow 0.625$

IGCSE Mathematics Revision Guide

Converting: percentages → fractions

Since *per cent* means 'out of 100', you convert a percentage to a fraction by taking the **percentage** as the numerator and writing the denominator as 100.

For example, 45% → 45% = $\frac{45}{100}$ which then simplifies to $\frac{9}{20}$.

Converting: percentages → decimals

Divide the percentage number by 100.

For example, 73% → 73% = 73 ÷ 100 = 0.73

Converting: decimals → percentages

Multiply the decimal number by 100.

For example, 0.84 → 0.84 = 0.84 × 100 = 84%

Converting: fractions → percentages

Divide the numerator by the denominator and then multiply by 100.

$\frac{5}{8}$ → 5 ÷ 8 × 100 = 0.625

Calculating a percentage

To calculate a percentage of a quantity, multiply the quantity by the percentage. You can write the percentage as either a decimal or a fraction. When working with a calculator it is usually easier to use a decimal, which is referred to as the **multiplier**.

For example, to find 26% of $130 → 26% = 0.26,
so 130 × 0.26 = 33.8 → $33.80 (0.26 is the multiplier.)

> **Revision tip**
>
> It is very useful to have quick recall of certain facts. Try to learn the percentage and decimal equivalents of $\frac{1}{2}, \frac{1}{4}, \frac{3}{4}, \frac{1}{8}, \frac{1}{10}, \frac{1}{5}, \frac{1}{3}, \frac{2}{3}$.
>
> $\frac{1}{2} = 0.5$
> $\frac{1}{4} = 0.25$
> $\frac{3}{4} = 0.75$
> $\frac{1}{8} = 0.125$
> $\frac{1}{10} = 0.1$
> $\frac{1}{5} = 0.2$
> $\frac{1}{3} = 0.333... = 0.33$ to 2 s.f.
> $\frac{2}{3} = 0.666... \approx 0.67$, to 2 s.f.

Percentage increases and decreases

Increase 240 kg by 15%

Method 1
Find 15% and then add it to 240.
240 × 0.15 = 36
240 + 36 = 276 kg

Method 2
Use a multiplier. Adding 15% to the original 100% will result in 115% of the original.
115% = 1.15
240 × 1.15 = 276 kg

Decrease 800 litres by 35%

Method 1
Find 35% and then subtract it from 800.
800 × 0.35 = 280
800 − 280 = 520 litres

Method 2
Use a multiplier. Subtracting 35% from the original 100% will result in 65% of the original.
65% = 0.65
800 × 0.65 = 520 litres

> **Revision tip**
>
> When working on a question involving money always answer with 2 **decimal places**, and remember to use the currency symbol.

Percentage change

The change could be either an increase or a decrease, but the calculation is identical:

$$\text{Percentage change} = \frac{\text{change}}{\text{original amount}} \times 100$$

For example, Adam sells his car for €4000. He had originally bought it for €4995. What is his **percentage loss** (to 2 s.f.)?

$$\text{Percentage loss} = \frac{4995 - 4000}{4995} \times 100 = 19.9\dot{1}\dot{9} \approx 20\%$$

Quick test

1. Which of these fractions are equivalent?

   $\frac{4}{7}$ $\frac{8}{15}$ $\frac{12}{21}$ $\frac{16}{28}$ $\frac{18}{36}$ $\frac{24}{42}$

2. Complete this table.

Decimal	Fraction	Percentage
0.875		
	$\frac{5}{16}$	
		15%

3. a) A stretch of motorway measuring 235 km is extended by 12%. How long, in kilometres, is the motorway now?

   b) Aarav reduces his loan of $3500 by 15%. How much does he owe now?

4. Airline fares are increased. A ticket from New York to Moscow used to cost $580.50 but is now $615.33

   What is the percentage increase?

Fractions and percentages 2

Extended

When a fraction is converted to a decimal, the decimal may work out exactly, for example: $\frac{7}{8} = 0.875$. This is a **terminating decimal**.

Some fractions convert to decimals that have a repeating pattern. For example:

- $\frac{1}{3} = 0.333... = 0.\dot{3}$
- $\frac{7}{11} = 0.63636363... = 0.\dot{6}\dot{3}$
- $\frac{8}{55} = 0.1454545... = 0.1\dot{4}\dot{5}$
- $\frac{41}{333} = 0.123123123... = 0.\dot{1}2\dot{3}$

These are called **recurring decimals**. They are made up of a repeating pattern in the decimal, identified by the dots above the digits.

Changing a recurring decimal to a fraction

To change 0.1454545... to a fraction, let $x = 0.1\dot{4}\dot{5}$

You need to shift the recurring pattern to the left-hand side of the equals sign.

To do this, multiply by 1000 to get $145.\dot{4}\dot{5}$, so $1000x = 145.\dot{4}\dot{5}$

In the same way, $10x = 1.\dot{4}\dot{5}$

Subtracting.... $1000x - 10x = 990x$

and $\quad 145.\dot{4}\dot{5} - 1.\dot{4}\dot{5} = 144$

so $\quad 990x = 144$

and $\quad x = \frac{144}{990}$ which simplifies to $\frac{8}{55}$

Revision tip

A recurring decimal will always convert to a fraction.

Simple and compound interest

Interest can be paid to you by a bank or investment firm when you save money with them. There are two methods of calculating interest: **simple interest** and **compound interest**.

For example, Kai invests $20 000 for 3 years at 1.5% simple interest.

Each year Kai will earn $\frac{20\,000 \times 1.5}{100} = \$300$ in interest, so after 3 years he will have an additional $900.

Alternatively, Kai could go to a bank paying compound interest. This means that the $300 he earns after the first year is added onto his initial amount, $20 300, and this is the amount used to calculate the interest for the second year, $\frac{20\,300 \times 1.5}{100} = \$304.50$

This added to the investment and the calculation for the third year is $\frac{20\,604.50 \times 1.5}{100} = \$309.0675$, or $309.07

So Kai would earn $913.57 over the 3 years.

In these calculations the original amount of money is called the **principle**.

Revision tip

Interest is an extra amount of money that is paid to someone in return for having use of money provided by them.

Compound interest formula

In the example about compound interest on page 13, you could have used a multiplier of 1.015 to find the new value of the investment amount each year.

The calculation would be:

$20\,000 \times 1.015 \times 1.015 \times 1.015 = \$20\,000 \times 1.015^3 = \$20\,913.57$, showing an increase of $913.57.

As a **formula**: final amount = principal amount $\left(1 + \dfrac{\text{interest rate}}{100}\right)^{\text{years}}$

Final amount = $\$20\,000 \left(1 + \dfrac{1.5}{100}\right)^3 = \$20\,913.57$

> **Revision tip**
>
> If a multiplier of less than 1 is used in the compound interest formula, it produces a decreasing amount – called compound decay. Depreciation could be calculated like this.

Extended

Reverse percentages

When you use reverse percentages, you start with a final amount and work backwards to find the starting amount.

> **Example**
>
> After three years with a football club a player is sold for $2.07 million. This is an increase of 15% on the price the club originally paid for him. How much did they pay three years ago?
>
> **Unitary method:** $2\,070\,000 represents 115%, so 1% would be $\dfrac{\$2\,070\,000}{115} = \$18\,000$
>
> Multiply 18 000 by 100 to find 100% (the starting value). $18\,000 \times 100 = \$1\,800\,000$ or $1.8 million.
>
> **Multiplier method:** an increase of 15% gives a multiplier of 1.15
>
> Divide the final amount by this: $\dfrac{\$2\,070\,000}{1.15} = \$1\,800\,000$ or $1.8 million.
>
> This method also works for decreasing values.

> **Example**
>
> A house bought five years ago has been sold for $460 000. This represents a drop in value of 20%. What was the value of the house five years ago?
>
> **Unitary method:** a drop of 20% means its current value is 80% of what it was.
>
> Original value = $\dfrac{\$460\,000}{80} \times 100 = \$5750 \times 100 = \$575\,000$
>
> **Multiplier method:** a drop of 20% gives a multiplier of 0.8
>
> Original value = $\dfrac{\$460\,000}{0.8} = \$575\,000$

Quick test

1. Which of these fractions produce recurring decimals?

    $\dfrac{8}{15}$ $\dfrac{9}{16}$ $\dfrac{4}{7}$ $\dfrac{2}{5}$ $\dfrac{3}{8}$ $\dfrac{5}{12}$

2. Convert $0.23\dot{6}$ to a fraction in its simplest terms.
3. Chelsey lends Kelsey $4000 for 5 years. Kelsey will pay 2.5% simple interest.
    How much interest does Chelsey receive from Kelsey?
4. Erik has some money to invest. He finds two competing offers, each for 8 years.
    Plan A requires €10 000 and pays 2.25% simple interest.
    Plan B requires €10 500 and pays 1.65% compound interest.
    Which plan gives Erik the better return for his money?

Extended

5. A small boat is sold for $7345, making a loss of 35% for its owners. What was its original value?

The four rules

Order of operations

Imagine you had to complete the calculation: $3^2 + 4 \times 2 - 6 \div 3$.

Ian says the answer is $-\frac{52}{3}$, Zara says the answer is 24 and Gail says the answer is 15. Who is right? There needs to be an agreed order for carrying out operations so that everyone who does the calculation arrives at the same answer, every time.

The correct answer is 15. Gail has applied the **BIDMAS** (or **BODMAS**) rule, which gives the order for carrying out operations:

B	**B**rackets ()
I/O	**I**ndices or p**O**wers
D	**D**ivide
M	**M**ultiply
A	**A**dd
S	**S**ubtract

> **Revision tip**
>
> You must follow BIDMAS (or BODMAS), as it gives the order in which the operations must be carried out.

So Gail correctly identified that 3^2 is done first and then 4×2 and $6 \div 3$ need to be done, leaving $9 + 8 - 2 = 15$

For Ian to be correct some brackets would need to be introduced:
$(3^2 + 4) \times (2 - 6) \div 3 = -\frac{52}{3}$

And for Zara: $(3^2 + 4) \times 2 - 6 \div 3 = 24$.

Adding and subtracting fractions

Before fractions can be added or subtracted they must have a common denominator. This involves changing one or both fractions to equivalent fractions, so that they both have the same denominator. This common denominator is the lowest common multiple of the original denominators.

> **Example**
>
> a) Find $\frac{7}{10} + \frac{1}{6}$
>
> The lowest common multiple of 10 and 6 is 30, so form two equivalent fractions:
>
> $\frac{3 \times 7}{3 \times 10} + \frac{5 \times 1}{5 \times 6} = \frac{21}{30} + \frac{5}{30} = \frac{26}{30}$, which simplifies to $\frac{13}{15}$
>
> b) Find $5\frac{7}{8} - 3\frac{2}{3}$
>
> Split the calculation into: $\left(5 + \frac{7}{8}\right) - \left(3 + \frac{2}{3}\right)$
>
> $= 5 - 3 + \left(\frac{7}{8} - \frac{2}{3}\right)$
>
> The fractions have a common denominator of 24.
>
> $= 5 - 3 + \left(\frac{21}{24} - \frac{16}{24}\right)$
>
> $= 2\frac{5}{24}$

IGCSE Mathematics Revision Guide

Multiplying and dividing fractions

To multiply fractions, you multiply the numerators together and multiply the denominators together. If any of the fractions are mixed numbers, change them to improper fractions before you start multiplying.

Example

a) Find $\dfrac{3}{5} \times \dfrac{2}{9}$

   Method 1: $\dfrac{3 \times 2}{5 \times 9} = \dfrac{6}{45} = \dfrac{2}{15}$

   Method 2: $\dfrac{\cancel{3}^1 \times 2}{5 \times \cancel{9}_3} = \dfrac{2}{15}$ Simplify by cancelling before multiplying.

b) Find $2\dfrac{3}{4} \times 4\dfrac{4}{5}$

   $2\dfrac{3}{4} \times 4\dfrac{4}{5} = \dfrac{11}{\cancel{4}_1} \times \dfrac{\cancel{24}^6}{5} = \dfrac{66}{5} = 13\dfrac{1}{5}$

To divide one fraction by another, turn the second fraction upside down (forming its **reciprocal**) and then multiply the fractions together.

Example

Find $2\dfrac{3}{4} \div \dfrac{5}{12}$

$2\dfrac{3}{4} \div \dfrac{5}{12} = \dfrac{11}{\cancel{4}_1} \times \dfrac{\cancel{12}^3}{5} = \dfrac{33}{5} = 6\dfrac{3}{5}$

Finding a fraction of a quantity

This involves multiplying the quantity by the fraction. Remember that to make a whole number into a fraction, you put it over a denominator of 1, and that in mathematics the word 'of' means multiply.

Example

Find $\dfrac{7}{8}$ of 344 kg.

$\dfrac{7}{8} \times 344 = \dfrac{7}{\cancel{8}_1} \times \dfrac{\cancel{344}^{43}}{1} = 301\,\text{kg}$

> **Revision tip**
>
> With word problems such as questions 2 and 4, you need to work out the correct operation to use. Reading the question aloud often helps you identify the right operation to use.

Quick test

1. Insert ×, ÷, +, − or ( ) where necessary to make each calculation true.

   a) 5 2^2 3 = 3

   b) 3^3 6   2 = 108

2. A zoo charges $15.50 per adult and $8.25 per child. A family ticket (2 adults and 2 children) costs $40. How much cheaper is it to buy a family ticket instead of 2 adult and 2 child tickets?

3. Work out each of these, leaving your answer in its simplest form.

   a) $2\dfrac{3}{5} + 4\dfrac{1}{8}$

   b) $2\dfrac{2}{5} - 1\dfrac{3}{8}$

   c) $2\dfrac{4}{5} \times 3\dfrac{5}{8}$

   d) $4\dfrac{4}{5} \div 1\dfrac{1}{15}$

4. Gunawan has to pay a $\dfrac{3}{20}$ deposit on a $1500 holiday. How much deposit does he pay?

Directed numbers

Everyday use of directed numbers

Positive and negative numbers together make up the directed numbers. There are many everyday situations in which directed numbers are used.

- Thermometers are good visual aids to help you remember that numbers below or to the left of zero are negative and numbers above or to the right of zero are positive.
- If it was −4 °C at midnight and 12 °C at midday, the temperature has risen by 16 °C.
- Many high-rise office blocks have underground parking. If you park on floor −3 and take the lift to your office on floor 15, you travel 18 floors.

The number line

Using a number line can help you to decide whether one number is bigger or smaller than another. When two different numbers are shown on a number line the smaller of the two is the one furthest to the left.

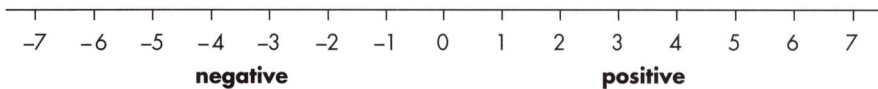

You can see from the number line that −5 is smaller than 1, because −5 is to the left of 1

This can be written as an **inequality**: −5 < 1

6 is bigger than −2 because 6 is to the right of −2

This can be written as an inequality: 6 > −2

Revision tip

Make sure you understand the inequality signs: < means 'is less than', > means 'is greater than', ⩽ means 'less than or equal to' and ⩾ means 'greater than or equal to'.

Adding and subtracting numbers

Adding a positive number to another number: on a number line, the answer will be to the right of the original number. −6 + 9 = 3

Adding a negative number to another number: on a number line, the answer will be to the left of the original number. 4 + (−7) = −3

Subtracting a positive number from another number: on a number line, the answer will be to the left of the original number. 3 − (+8) = −5

Subtracting a negative number from another number: on a number line, the answer will be to the right of the original number. −2 − (−6) = +4

Multiplying and dividing directed numbers

First, multiply or divide the numbers, ignoring their signs. To work out whether the answer is positive or negative, remember the rule:

If the signs of the two numbers are the same, the answer is positive; if the signs of the two numbers are different, the answer is negative.

For example: −3 × 6 = −18   2 × 5 = 10   −8 ÷ 2 = −4   −15 ÷ (−5) = 3

Quick test

1. Write down the answers.
   a) −5 + 6
   b) −2 − (−8)
   c) 4 − (10 ÷ (−2))
   d) −3 + (−2 × −6)

2. Put these calculations in order of their answers, lowest to highest.
   6 × (4 ÷ (−2))
   −4 + 10 − (−7) − 5
   5 × (36 ÷ (5 − (−7)))
   30 ÷ (−2 × (−3))

IGCSE Mathematics Revision Guide

Squares and cubes

Squares and square roots

On page 6 you were introduced to square numbers and square roots.

The square of a number is that number multiplied by itself, for example, the square of 7 is 49 as $7 \times 7 = 49$, which you write as 7^2 in **index notation**. When reading this, say, '7 squared'.

Another way to describe this is to say that 7 is the **square root** of 49, written as $\sqrt{49} = 7$.

Remember that a square root has two possible answers: -7×-7 is also $+49$, so $\sqrt{49} = \pm 7$, even though your calculator will only show the positive root, $+7$.

Revision tip

Make sure you know how to use the x^2 and $\sqrt{x}$ keys on your calculator.

Cubes and cube roots

The cube of a number is that number multiplied by itself twice, for example, $5 \times 5 \times 5 = 125$, which you write as 5^3 in index notation. When reading this, say, '5 cubed'.

Another way to describe this is to say that 5 is the **cube root** of 125, written as: $\sqrt[3]{125} = 5$.

Revision tip

Make sure you know how to use the x^3 and $\sqrt[3]{x}$ keys on your calculator.

Other powers and roots

Index numbers can be any number, not just ^{2} (square) and ^{3} (cube). An index number of ^{4} means that the number is multiplied by itself 4 times, so $5^4 = 5 \times 5 \times 5 \times 5 = 625$.

In the same way, $2^7 = 2 \times 2 \times 2 \times 2 \times 2 \times 2 \times 2 = 128$. You would say, 2 to the power 7.

Roots can also be other than square roots and cube roots. For example $\sqrt[5]{243} = 3$. You would say, the fifth root of 243.

When you write a number using index notation, such as 4^6, the number '4' is the base number and '6' is the index number.

Revision tip

Make sure you know how to use the x^y and $\sqrt[y]{x}$ buttons on your calculator.

Extended

Exponential growth and decay

You have already met the idea of **exponential growth** when calculating compound interest. The formula, as shown on page 14, can be applied to any situation where growth is described in terms of a multiplier that is > 1 and has a positive **index** number.

For example, suppose Erika invests $20 000 for five years at a compound interest rate of 1.75%
The multiplier is $1 + 0.0175 = 1.0175$, so Erika's final amount $= 20\,000 \times 1.0175^5 = \$21\,812.33$

Exponential decay will have a multiplier of less than 1.

For example, suppose the value of Kelvin's jet-ski has decreased by 15% each year he has owned it. He bought it for $8500 six years ago. What is its value now?
The multiplier is $1 - 0.15 = 0.85$, so the current value $= 8500 \times 0.85^6 = \$3205.77$

Quick test

1. Evaluate these expressions.

   a) 19^3
   b) $\sqrt[3]{12167}$
   c) $4^3 + \sqrt{3136} + 19^2 - \sqrt[3]{15\,625}$
   d) 6.5^3

2. Here are four numbers. Which is the odd one out? Why?

   $8^2 \quad \sqrt{4096} \quad 4^3 \quad \sqrt[3]{32\,768}$

3. a) Which is larger, 5^6 or 6^5? By how much?

   b) Which is smaller, $\sqrt[8]{6561}$ or $\sqrt[6]{4096}$? By how much?

Extended

4. In 1984 the number of nesting sites for the leatherback turtle was estimated to be 6500. Over the next 11 years there was a decline in nesting sites of 20% per year. What was the estimated number of nesting sites in 1995?

5. The popularity in running marathons has grown by approximately 3.3% per year since 2000. In 2000 the number of runners who completed a marathon in the USA was 350 000. What was the approximate number of runners completing a USA marathon in 2014? Give your answer to four significant figures (4 s.f.).

Ordering and set notation

Inequalities

To answer questions about **sets**, you need to know the meaning of these symbols:

=	is equal to
≠	does not equal
<	is less than
>	is greater than
⩽	is less than or equal to
⩾	is greater than or equal to

You need to be able to both place them between values to make an inequality true, and to list the possible values expressed by an inequality.

Example

a) Which inequality symbol links this pair of values?

44% of 250 tonnes ☐ $\frac{5}{6}$ of 130 tonnes

$\frac{44}{100} \times 250 = 110$ and $\frac{5}{6}$ of $130 = 108.\dot{3}$

So, 44% of 250 tonnes > $\frac{5}{6}$ of 130 tonnes

b) If $\sqrt{30} < n \leq 3.2^2$, what are the possible integer values of n?

$\sqrt{30} = 5.47...$ and $3.2^2 = 10.24$, so the possible values for n are $n = 6, 7, 8, 9, 10$.

Sets

A set is a collection of items, called **elements**. Sets are usually identified by a capital letter with the elements of the set enclosed inside curly brackets, { }.

Sometimes you may not be able to list the full set of elements, such as the set containing all the prime numbers. In these cases, use a series of dots to indicate that there are more elements than could be listed.

$F = \{1, 2, 3, 4, 6, 8, 12, 24\}$

$Y = \{$January, February, March, ...$\}$

{prime numbers} = $\{2, 3, 5, 7, 11, ...\}$

The number of elements in set Y is 12, written as $n(Y) = 12$.

The set that contains all the elements of a number of subsets is called the **universal set** and has the symbol ε.

If the set, ε, contains the positive numbers 1–10, then:

$ε = \{1, 2, 3, 4, 5, 6, 7, 8, 9, 10\}$

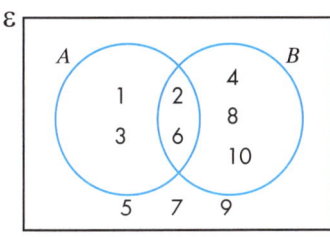

Two of the subsets of ε are:

$A = \{1, 2, 3, 6\}$ and $B = \{2, 4, 6, 8, 10\}$

This information can be shown in a **Venn diagram**.

Each element of ε is in A, B, both or neither.

$A \cap B$ is the **intersection** of A and B, those elements that are in both A and B: $A \cap B = \{2, 6\}$

$A \cup B$ is the **union** of A and B, those elements that are in A or B or both: $A \cup B = \{1, 2, 3, 4, 6, 8, 10\}$

> **Revision tip**
>
> When you draw any Venn diagram ensure that every element of the universal set is included somewhere within the rectangle.

Extended

To identify an element of a set, you use the symbol $\in$.

To specify that 3 is an element of set A write: $3 \in A$

You can also write $7 \notin A$, which means that 7 is *not* an element of set A.

A' (sometimes read 'not A') is the complement of A: $A' = \{4, 5, 7, 8, 9, 10\}$

The three symbols can be used in combination.

$(A \cap B)' = \{1, 3, 4, 5, 7, 8, 9, 10\}$ and $A' \cap B = \{4, 8, 10\}$

> **Revision tip**
>
> Make sure you can recognise the meaning of the symbols: $\in, \notin, \emptyset, \subset, \subseteq$

If the set of pets, P, owned by Nuria is $P = \{\text{cat, parrot, rabbit}\}$, then P has 8 **subsets**, which are smaller sets that can be made from the elements of the original set:

- {cat, parrot}, {cat, rabbit}, {parrot, rabbit} each with two elements
- {cat}, {parrot}, {rabbit} each with 1 element
- The *empty* set, $\emptyset$ (no elements) and the set P itself
- {cat, fish} is not a subset because it has an element (fish) which is not in the set P.

If set $Q = \{\text{cat, parrot}\}$ then Q is a subset of P because every element of Q is in P. You can write this as: $Q \subseteq P$.

All subsets except P are called proper subsets. The symbol for a **proper subset** is $\subset$. So, {rabbit} $\subset P$, but $P \not\subset P$.

All subsets, except the original set P, are proper subsets.

Quick test

1. Put the correct inequality symbol between the values in each pair.

   a) $\sqrt[3]{75}$ ☐ 2.05^2 b) 0.2 litres ☐ 200 cl   c) $\frac{3}{8}$ of 200 km ☐ $\frac{2}{5}$ of 185 km

2. List all the possible integer values for n if $-6 \leqslant n < 1.6^2$

3. a) List the elements of:

      i) $A \cap B$ ii) $A \cup B$

   b) Find $n(A \cap B)$

   c) Find: $n(B \cap C)$.

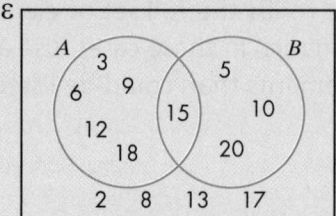

   Extended

   d) List the elements of:

      i) $(A \cup B)'$ ii) $A' \cup B$

   e) Find: $n(A \cap B)'$.

   f) Redraw the diagram to show the subset $C = \{\text{multiples of 6}\}$.

Ratio, proportion and rate 1

Ratio

A **ratio** is a way of comparing the sizes of two or more quantities. A colon (:) is used to show ratios.

For example, if 10 ml of fruit cordial is to be mixed with 250 ml water, the mix of cordial to water is 10 : 250, which simplifies to 1 : 25

You can simplify ratios if you can divide each value by a common factor.

The order that the values are written is important – they follow the same order as the description.

It is important to use the same working units for all values. If the mix of cordial to water had been given as 10 ml to 0.25 litres (10 ml : 0.25 litres) you would need to change 10 ml to 0.01 litres (giving a ratio of 0.01 : 0.25, and simplifying to 1 : 25), or to change 0.25 litres to 250 ml before writing the ratio as 10 : 250. When the units are the same for all the values, the ratio has no units.

Ratios can be written as fractions.

If the ratio of blond to brunette students in a class is 2 : 3, it means that for every 5 students (adding blond and brunette values together), 2 will be blond and 3 will be brunette.

So $\frac{2}{5}$ of the class is blond and $\frac{3}{5}$ is brunette.

> **Revision tip**
>
> You cannot simplify a ratio until every part is expressed in the same units.

Dividing amounts in a given ratio

To divide an amount into a given ratio, add the given values together to see how many parts or shares there are in total. Each value in the ratio can be expressed as a fraction of this total.

> **Example**
>
> Alan, Beth and Chris share 60 DVDs in the ratio 5 : 6 : 4.
>
> There are 15 'parts' (5 + 6 + 4 = 15).
>
> So Alan has $\frac{5}{15} \times 60 = 20$ DVDs, Beth $\frac{6}{15} \times 60 = 24$ DVDs, and Chris $\frac{4}{15} \times 60 = 16$ DVDs.

The reverse process can also be used to find missing information.

> **Example**
>
> Daria and Erik shared company profits in the ratio 5 : 4.
>
> If Daria received $2500 how much did Erik receive?
>
> Daria's share was $\frac{5}{9}$ of the profit, so $\frac{1}{9}$ of the profit was $\frac{\$2500}{5} = \$500$.
>
> Erik's share was $\frac{4}{9}$, so $4 \times 500 = \$2000$

Extended

Increases and decreases using ratios

Suppose a travel company estimates that the baggage for 60 people on one of its holidays will total 2100 kg. If five extra people are booked on the holiday, what will the total baggage estimate increase to?

Increase 2100 in the ratio 65 : 60 (or 13 : 12) and find $\frac{13}{12} \times 2100 = 2275$ kg

If 10 of the 60 people cancel, decrease 2100 in the ratio 50 : 60, or 5 : 6 and find $\frac{5}{6} \times 2100 = 1750$ kg

Map scales

The scales used on maps are written as ratios in the form $1 : n$ (referred to as unitary form).

Suppose a map of Africa has a scale of 1 : 8 000 000. The distance on the map from Cairo to Khartoum is 20 cm. How far is the actual distance?

The scale tells you that 1 cm on the map = 8 000 000 cm on the ground.
8 000 000 cm = 80 km, so the actual distance is $20 \times 80 = 1600$ km.

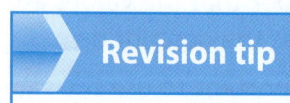

Revision tip

1 km = 1000 m = 100 000 cm

Speed

The relationship between speed (S), time (T) and distance (D) is illustrated in this diagram.

For example, Jon cycled for 4 hours at an **average speed** of 15 km/h. How far did he travel?

distance = speed × time, $15 \times 4 = 60$ km

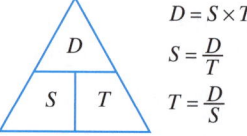

$D = S \times T$
$S = \frac{D}{T}$
$T = \frac{D}{S}$

Speed can also be used to describe the rate at which a quantity changes.

> **Example**
>
> A pond that holds 1500 litres of water is being filled through a pipe with water flowing at 20 litres a minute. How long will it take to fill the pond?
>
> $1500 \div 20 = 75$ minutes $\div 60 = 1.25$ hrs = 1 hour 15 minutes

Quick test

1. Simplify these ratios.
   a) 24 : 60
   b) 2 litres : 600 ml
   c) 25 m : 40 mm
   d) 3 months : 6 years 9 months
2. A bunch of flowers has lilies and carnations in the ratio 5 : 3. What fraction of the bunch is carnations?
3. In a marathon race 1560 runners completed the course. The ratio of those finishing to those who did not was 52 : 3. How many runners did not finish?
4. Monika has a map with a scale of 1 : 20 000. On the map, a castle is 6 cm from a beach where Monika is standing. How far away from Monika is the actual castle?

Extended

5. A $5 banknote is 120 mm long. A $500 banknote is longer, in the ratio 4 : 3. How long, in millimetres, is the $500 banknote?
6. Rana and Lena are travelling 110 km to the coast. Rana takes the 9:30 am bus, which travels at an average speed of 40 km/h. Lena will use her car and drive at an average speed of 60 km/h. If Lena wants to arrive at the same time as Rana, what time should she set off?

Ratio, proportion and rate 2

Direct proportion

If two connected values are either both increasing or both decreasing at a constant rate they are in **direct proportion**. A good example of this is currency conversion.

For example, Deepa had £150 to change into Indian rupees (₹). She received ₹12 945. These values are in direct proportion because if she had £300 Deepa would expect ₹25 890.

The best way to solve a direct proportion problem is to find the single unit value, in this case the exchange rate is £1 = ₹86.3. This is called the unitary method.

The graph of a direct proportion relationship is a straight line with a positive **gradient** passing through (0,0).

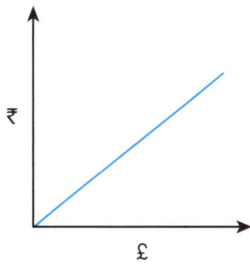

Inverse proportion

This is the opposite of direct proportion. As one value increases the other decreases. A good example of this is average speed over a fixed distance.

As speed increases, the time taken for the journey decreases, and as speed decreases the time taken increases. For example, a journey of 200 km made at a speed of 40 km/h will take 5 hours. If the speed was 100 km/h then it would take only 2 hours.

In these calculations you need to work out the constant value. In the above example the distance travelled is fixed.

A graph of inverse proportion is a concave curve.

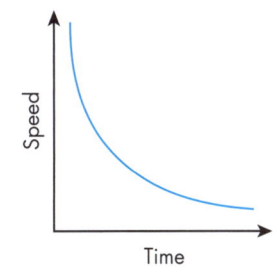

> **Revision tip**
>
> Memorise the shape of the graphs for direct and inverse proportion because they will help you remember the way that the two data sets are related in each case.

Density

The relationship between density (D), mass (M) and volume (V) is illustrated in this diagram.

For example: a small statue is made of bronze. The volume of the statue is 1120 cm³ and the density of the bronze used is 8.6 g/cm³.

What is the mass, in kg, of the statue?

Mass = density × volume → 8.6 × 1120 = 9632 g = 9.632 kg

> **Revision tip**
>
> 1 kg = 1000 g
> 1000 kg = 1 tonne

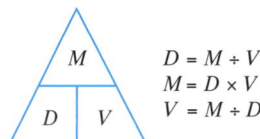

$D = M \div V$
$M = D \times V$
$V = M \div D$

Pressure

The relationship between pressure (P), force (F) and area (A) is illustrated in this diagram.

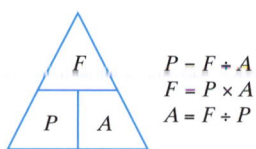

Pressure is the force per unit area and is expressed in newtons (N) per square metre (m²). A force of 1 N applied to 1 m² is called 1 pascal (Pa).

Gravity affects the downward force of an object. Normally g = 9.81 m/s² so a mass of 1 kg exerts a force of 9.81 N. To make calculations easier take g = 10 m/s²

For example: a metal safe has a mass of 80 kg and exerts a force of 2000 Pa on the ground. What is the base area of the safe, in cm²?

Force = 80 × 10 = 800 N

Area = force ÷ pressure → 800 ÷ 2000 = 0.4 m² = 4000 cm²

Quick test

1. 6 litres of paint cost $21.60
   How much will 14 litres cost? How many complete litres can Anaya buy for $30?
2. 20 cupcakes cost $17. Sue has $10. How many cupcakes can she buy?
3. Paulo types an assignment at an average of 18 words per line and it takes 160 lines. He changes the margin sizes and the font so that he fits 24 words per line. How many lines will the assignment take now?
4. A label for a bottle of lemonade is 9 cm high and 15 cm wide. The label is redesigned to wrap around more of the bottle and is now 20 cm wide. The area is the same. What is the height of the new label?
5. a) A glass paperweight has a volume of 80 cm³ and has a mass of 200 g. What is the density of the glass used?
   b) 3.25 kg of metal A, with a density of 2.6 g/cm³ is mixed with 1200 cm³ of metal B which has a mass of 2040 g. What is the density of the resulting metal, to 2 d.p.?
6. Malek has a mass of 50 kg and the contact patch that the tyres of his bicycle make with the road are 1020 mm² for each tyre. Malek's bicycle has a mass of 15 kg. Assuming that Malek is cycling on a length of road that is horizontal, what pressure does Malek exert on the road (to 2 s.f.)?

Estimation and limits of accuracy 1

Rounding whole numbers

At the 2016 Rio Olympics, 4924 medals were given out.

What is 4924 to the nearest 10? The required accuracy is to the nearest 10, so the digit to examine is in the units column (4). Because 4 < 5, **round** the number down to 4920. If the units digit had been ⩾ 5 you would have rounded up to 4930.

To round to the nearest 1000 you examine the digit in the hundreds column. For 4924 it is 9 (⩾ 5), so you round up to 5000.

The general rule when **rounding** whole numbers is to decide where the cut-off point is, for example, the required number of **significant figures**. Then examine the next digit to the right. When this digit is < 5, round down, if it is ⩾ 5 round up by adding 1 to the digit in the required position. Then replace all the digits to the right of this by 0.

Rounding decimals (d.p.)

The idea is the same as for whole numbers, but for decimals you just remove all the digits after the one to which you have rounded.

For example, the first 10 digits of **pi (π)** are 3.141 592 653… Written to two **decimal places** (2 d.p.) this is 3.14, and to four decimal places it is 3.1416

Significant figures (s.f.)

The most significant figure in any number is the left-most, non-zero digit.

For example, the value of the acceleration due to gravity is 9.806 65 m/s^2. Written to four significant figures (4 s.f.) this is 9.807, and to two significant figures it is 9.8 m/s^2.

A number such as 0.010084 may be written as 0.010 (2 s.f.) and 0.0101 (3 s.f.).

Rounding in context

Sometimes you will need to round answers to a reasonable degree of accuracy. In these cases, the context of the problem will give an indication of the accuracy required. Generally, an answer need not have a greater degree of accuracy than the values that have been used in the calculation. Questions involving trigonometrical ratios or π are going to present answers with many decimal places. So, for example, if the question involves values written to 1 and 2 d.p. it is reasonable to round your answer to 2 d.p. If a question involves calculating a number of people or objects then it would be usual to answer to the nearest whole unit. Think carefully about what the answer is conveying to the reader – just how much accuracy is helpful?

> **Revision tip**
>
> A common error is to confuse rounding to decimal places and rounding to significant figures. Make sure you remember the difference.

Quick test

1. Round each number to the number of decimal places (d.p.) indicated.
   a) 23.764 (2 d.p.)
   b) 21.276 (1 d.p.)
   c) 34.897 (2 d.p.)
   d) 0.0562 (3 d.p.)
2. The number of spectators at a tennis match was reported by one newspaper as 7300 (to the nearest 100) and 7250 in a magazine (to the nearest 10). What is the largest and smallest possible number of spectators at the match?
3. Round each number to the number of significant figures (s.f.) indicated.
   a) 23.764 (3 s.f.)
   b) 21.276 (1 s.f.)
   c) 34.897 (4 s.f.)
   d) 0.0562 (1 s.f.)

Estimation and limits of accuracy 2

Upper and lower bounds

Any recorded measurement has been rounded to some degree of accuracy. The height T of a tree, measured as 9 metres to the nearest metre, could be anything between 8.5 m and 9.5 m.

8.5 rounds up to 9 and anything less than 9.5 will round down to 9.

You would express this as $8.5 \leq T < 9.5$.

These limits, 8.5 and 9.5, are the **lower bound** and **upper bound**. They are also called the **limits of accuracy**, and to find them you need to add or subtract half the measurement of accuracy to or from the measurement given.

For example, the mass S of a sack of coffee beans is 65 kg, measured to the nearest 500 g.

The measurement of accuracy is 500 g so the possible error is ± 250 g.

This means the limits of accuracy, given by the lower and upper bounds are:

$$64.75 \leq S < 65.25$$

> **Revision tip**
>
> The lower and upper bounds are always written with the inequality symbols in this order: lower bound $\leq$ the measurement $<$ upper bound.

Extended

Upper and lower bounds for calculation

You can use lower and upper bounds to find least and greatest possible values for the result of a calculation.

Suppose you are carpeting a room that has been measured, to the nearest 10 cm, as 8.4 m long and 6.5 m wide. Then the lower and upper bounds of each measurement will be 6.5 m ±0.05 m and 8.4 m ±0.05 m.

8.35 m $\leq$ length $<$ 8.45 m and 6.45 m $\leq$ width $<$ 6.55 m

The lower bound for the area is $8.35 \times 6.45 = 53.8575 \text{ m}^2 \approx 53.86 \text{ m}^2$

The upper bound for the area is $8.45 \times 6.55 = 55.3475 \text{ m}^2 \approx 55.35 \text{ m}^2$

The limits of accuracy for the area of the room are: $53.86 \text{ m}^2 \leq$ area $< 55.35 \text{ m}^2$

If the carpet chosen is $18 per square metre, the cost of carpeting the room could be between $969.48 and $996.30

You can also use lower and upper bounds in different combinations.

> **Example**
>
> A chef estimates that he uses 12.45 litres (to the nearest 60 ml) of honey for a particular dessert each week. Each dessert uses 53 ml (to the nearest 2 ml). What are the minimum and maximum number of desserts that he makes?
>
> The lower and upper bounds of the weekly total:
> 12 420 ml $\leq$ honey $<$ 12 480 ml
>
> The lower and upper bounds used for each dessert: 52 ml $\leq$ dessert $<$ 54 ml
>
> The upper bound of the number of desserts made is $\frac{12480}{52} = 240$
>
> The lower bound of the number of desserts made is $\frac{12420}{54} = 230$
>
> Calculations like this can also be used to find measures such as speed, rate of flow or fuel consumption.

Quick test

1. Joshua runs 100 m in 13.6 seconds, measured with a stopwatch with an accuracy of $\frac{1}{10}$ of a second. What is the fastest time that he could have taken?

   Rahim runs 100 m in 13.56 seconds, measured with a stopwatch with an accuracy of $\frac{1}{100}$ of a second. Can Rahim claim to be faster than Joshua?

2. A new hockey pitch is being marked out, measuring 91.5 m × 50.0 m. Each side is measured to the nearest 50 cm. What is the maximum length of each side?

Extended

3. Luggage checked in for a flight to Florida is weighed to the nearest 100 g, with a maximum allowed mass of 22 kg per passenger. There are 238 passengers. What is the upper bound for the total mass of the luggage?

4. The length of a metal bar is measured as 250 mm, to the nearest mm. It is heated in a blacksmith's furnace and expands to 257 mm, to the nearest mm. What is the greatest possible expansion of the bar?

5. A forest manager is working out the tree density of some woodland. The woodland is a rectangle, measured as 1000 m × 250 m (each to the nearest 10 m). The number of trees is 2450, to the nearest 50.

   What are the upper and lower bounds for the tree density? This is measured in trees/hectare, where 1 hectare = 10 000 m².

   If the minimum tree density for woodland to be declared a forest is 100 trees/hectare, could this woodland be a forest?

Standard form

Writing numbers in standard form

You have already met square and **cube numbers** where, in a number such as 5^2, there is an index number, ^{2}. This is also called a **power**, so instead of saying 'five squared', you could say, 'five to the power two'.

In **standard form**, numbers are written as a number between 1 and 10 multiplied by a power of 10.

The format is $A \times 10^n$ where $1 \leq A < 10$, and n is an integer.

Standard form is very useful when you are describing very big (a positive power of 10) or very small (a negative power of 10) numbers, such as the distance, in kilometres, from Earth to the Sun, or the width, in millimetres, of an atom.

Your calculator display will probably only show a number with up to 10 digits, before switching to display in standard form whenever 11 or more digits are required. It is important that you know how to enter a number in standard form and how to read a number that is displayed in standard form. Typically, you will use a button, and see on the screen, '$\times 10^n$'.

> **Revision tip**
>
> You can enter numbers on your calculator in standard form. Make sure you know how to do this on your calculator – not all work in the same way.

Standard form for numbers greater than 1

67	$= 6.7 \times 10$	$= 6.7 \times 10^1$
167	$= 1.67 \times 100$	$= 1.67 \times 10^2$
2167	$= 2.167 \times 1000$	$= 2.167 \times 10^3$

Standard form for numbers less than 1

0.67	$= 6.7 \times 0.1$	$= 6.7 \times 10^{-1}$
0.0167	$= 1.67 \times 0.01$	$= 1.67 \times 10^{-2}$
0.002\,167	$= 2.167 \times 0.001$	$= 2.167 \times 10^{-3}$

Calculating with standard form

> **Example**
>
> a) In 2013 an estimated 3.5 billion tweets a week were sent by 240 million active Twitter users. How many tweets on average would an active Twitter user have sent in the year?
>
> $3\,500\,000\,000 = 3.5 \times 10^9$, $240\,000\,000 = 2.4 \times 10^8$
>
> This means an average weekly tweet rate of:
>
> $(3.5 \times 10^9) \div (2.4 \times 10^8) = (3.5 \div 2.4) \times (10^9 \div 10^8) = 1.458 \times 10^1 = 14.58$
>
> In 2013 an active Twitter user would have sent approximately $14.58 \times 52 = 758$ tweets.

b) A sheet of aluminium foil measures $0.2\,m \times 0.2\,m \times 1.6 \times 10^{-5}\,m$.

The mass of the sheet is $1.728 \times 10^{-9}\,kg$. What is the **density** of aluminium, in g/cm^3?

density = mass ÷ volume

The volume of the sheet
$= 2 \times 10^{-1} \times 2 \times 10^{-1} \times 1.6 \times 10^{-5}$
$= 2 \times 2 \times 1.6 \times 10^{-1} \times 10^{-1} \times 10^{-5}$
$= 6.4 \times 10^{-7}$ cubic metres

The density of aluminium
$= (1.728 \times 10^{-9}) \div (6.4 \times 10^{-7})$
$= (1.728 \div 6.4) \times (10^{-9} \div 10^{-7})$
$= 0.27 \times 10^{-2}$
$= 2.7 \times 10^{-3}$ kg per cubic metre

Converting to g/cm^3, density = $2.7 \times 10^{-3} \times 1000 = 2.7\,g/cm^3$

Quick test

1. Write each number in standard form.
    a) 317   b) 31 700   c) 0.317   d) 0.000 317   e) 3.17
2. Write these as an ordinary decimal numbers.
    a) 4.87×10^5 b) 4.87×10^2 c) 4.87×10^{-2} d) 4.87×10^{-6}
3. Simplify these expressions, giving your answer in standard form.
    a) $(5 \times 10^3) \times (7.6 \times 10^2)$ b) $(2.8 \times 10^6) \times (7.6 \times 10^{-2})$ c) $(3.5 \times 10^{-4}) \times (7 \times 10^{-3})$
    d) $(6 \times 10^3) \div (7.5 \times 10^2)$ e) $(1.56 \times 10^{-4}) \div (2.5 \times 10^2)$ f) $(8.2 \times 10^{-2}) \div (4 \times 10^{-4})$
4. The spacecraft Juno became the fastest travelling manufactured object when it was pulled into Jupiter's orbit, reaching a speed of approximately $2.6 \times 10^5\,km/h$. Juno's journey had been a total of $2.8 \times 10^9\,km$ long. If Juno had travelled the entire distance at this top speed, how many days would the journey have taken?
5. A sheet of gold leaf is 1 millionth of 1 m thick. How many sheets of gold leaf would need to be laid on top of each other to make a pile as thick as a sheet of standard printer paper, with thickness of $9 \times 10^{-2}\,mm$?

Applying number and using calculators

Units of measurement

Through the centuries, countries have had their own units of measurement, but now the world has generally adopted an official standard measuring system – the metric system – which is used all over the world, except for the United States of America.

Unit	How to estimate it
Length	
1 metre	A long stride for an average person
1 kilometre	Two and a half times round a school track
1 centimetre	The distance across a fingernail
Mass	
1 gram	A small coin weighs a few grams
1 kilogram	A bag of sugar
1 tonne	A saloon car
Volume/capacity	
1 litre	A full carton of orange juice
1 centilitre	A small glass holds about 10 cl
1 millilitre	A full teaspoon is about 5 ml

Capacity is the volume (expressed in litres) of a liquid or a gas, for example, the volume of oil in a barrel is approximately 160 litres.

You need to be able to decide on the most appropriate units to use for a particular measurement. For example, you would measure the height of a tree in metres and not centimetres, and the mass of a bar of chocolate in grams and not tonnes.

> **Example**
> a) What units would be used to measure the length of the pencil you are using?
>
> Centimetres
>
> b) What would be the approximate dimensions, capacity and units of a children's paddling pool?
>
> A small circular pool might have a diameter of 1.5 m and a depth of water of 20 cm. Its capacity would be approximately 1500 litres.

This table gives the relationships and conversions between various units.

Length	Mass
10 **millimetres** = 1 **centimetre**	1000 **grams** = 1 **kilogram**
1000 millimetres = 100 centimetres = 1 **metre**	1000 kilograms = 1 **tonne**
1000 metres = 1 **kilometre**	

Capacity	Volume
10 **millilitres** = 1 **centilitre**	1000 litres = metre^3
1000 millilitres = 100 centilitres = 1 **litre**	1 millilitre = 1 centimetre^3

> **Revision tip**
>
> Ensure that you know the conversion between units. Remember that the prefixes milli- and kilo- refer to 1000.

Note the connection between capacity and volume: 1 litre = 1000 cm^3 and 1 ml = 1 cm^3.

Example

a) Change 1.45 m to mm.

1.45×1000

$1.45\,\text{m} = 1450\,\text{mm}$

b) Change 1.2 m^3 to litres.

$1.2 \times 1\,000\,000$

$1.2\,\text{m}^3 = 1\,200\,000\,\text{cm}^3 = 1\,200\,000\,\text{ml}$

$1\,200\,000 \div 1000$

$1\,200\,000\,\text{ml} = 1200\,\text{litres}$

Time

Time can be written in the 12-hour or the 24-hour clock.

In everyday life you normally use the 12-hour clock, which starts at midnight and runs through 12 hours to midday, then starts again and runs through 12 hours to midnight. Writing a.m. or p.m. after the time indicates whether it is before or after midday. You might have morning break at 10:30 a.m. and arrive home from school at 4:00 p.m.

The 24-hour clock uses four digits to represent the time. The first two give the hour and the last two give the minutes past the hour. You may catch the school bus at 07:45 and watch your favourite TV programme in the evening at 19:30.

Timetables are usually written in the 24-hour clock.

> **Example**
>
> a) Write 6:35 p.m. in the 24-hour clock.
>
>   6:35 p.m. = 18:35
>
> b) The Sapsan high-speed train leaves Moscow at 12:55 and arrives at St Petersburg at 17:05.
>
>   How long does the journey take?
>
>   4 hours and 10 minutes

Revision tip

Remember that an hour has 60 minutes and not 100 minutes, so 30 minutes is 0.5 hour and 0.75 hour is 45 minutes.

Currency conversions

To convert from one currency to another, you apply an exchange rate. These change continuously, depending on the world's financial markets.

> **Example**
>
> a) On 13 October 2016, €1 was worth = 7.44 Chinese yuan.
>
>   How many euros were 1000 yuan worth?
>
>   1000 ÷ 7.44 = 134.41 euros (to 2 d.p.)
>
> b) Entering Mexico from the USA, Ronald found that the exchange rate from US dollars to Mexican pesos was US$1 = 18.93 pesos. Ronald had $750 to change. How many pesos did he receive?
>
>   750 × 18.93 = 14 197.50 pesos
>
>   He received 14 197.50 pesos

Using a calculator efficiently

Being able to use a calculator efficiently is a vital skill.

Consider the calculation: $\frac{23.6 - 17.2}{1.25 + 2.63}$

If you keyed into your calculator: 23.6 − 17.2 ÷ 1.25 + 2.63 you would get 12.47, which is not the correct answer. Calculators follow the rules of BIDMAS, which tell you that ÷ will be done first, with the + and − done afterwards.

To ensure that the calculation is completed correctly you need to introduce brackets as you type it in:

This gives 1.65 (to 2 d.p.)

It is important that the values of the numerator and the denominator are worked out before the division takes place, in order to find the correct value.

Examples

a) Work out: $\dfrac{6^2 + \sqrt[3]{300}}{4.5 - 1.06}$ to 2 d.p.

The answer is: 12.41 (to 2 d.p.)

b) How would you enter $\dfrac{-4 + \sqrt{10^2 - 4 \times 4 \times -2}}{2 \times 4}$ into your calculator as one continuous sequence?

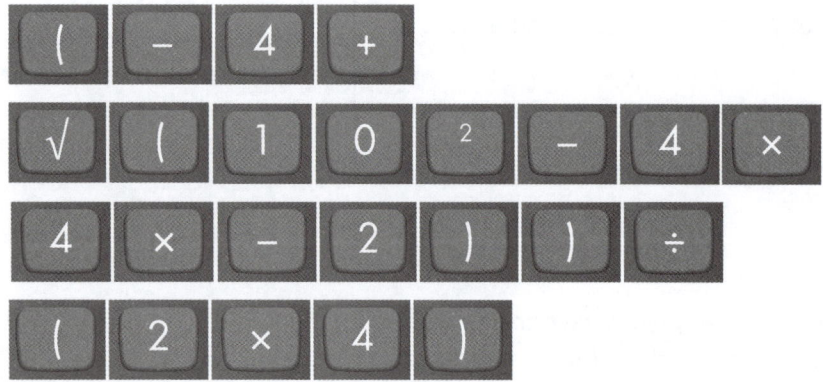

Alternatively, you could use the fraction button on your calculator and enter the numerator and denominator exactly as you see them.

> **Revision tip**
>
> Questions will often ask you to write down, as your answer, all the numbers on your calculator display. Make sure you do this, even if you have to write down 10 or 11 digits. After writing all the digits down, double check you have not missed any out or incorrectly written any twice.

Quick test

1. What metric unit would most likely be used to measure:
   a) the length of your foot   b) the mass of a car   c) the capacity of a teacup?

2. Convert the measurement given into the units shown in brackets.
   a) 230 mm (m)   b) 1.05 litres (ml)   c) 205 g (kg)   d) 350 cm³ (mm³)

3. The flight from Malaga to Copenhagen takes 3 hours 40 minutes. A flight arrives in Copenhagen at 14:10. What time did it depart from Malaga?

4. The Paris–Moscow express leaves Paris at 18:55 on a Wednesday and arrives in Moscow at 10:10 on Friday. If the time difference between Paris and Moscow is such that Moscow is 2 hours ahead of Paris, what is the elapsed time of the journey?

5. Vlad is travelling from Moscow to Copenhagen on a business trip. He changes 25 000 roubles into Danish krone. The exchange rate is 1 rouble = 0.11 krone.

   How many krone will Vlad get?

   On Vlad's return to Moscow the exchange rate has changed to 1 krone = 9.95 roubles. He gets 447.75 roubles. How many krone did he change (to the nearest whole krone)?

6. Use your calculator to find the value of each expression. Write down all the digits on your calculator display.

   a) $\dfrac{10.6^3 \times \sqrt[3]{120}}{2.45 + \cos 35°}$ b) $\dfrac{100 \div \sqrt{30}}{6.5^2 - 3.24^3}$

Exam-style practice questions

1
a) Write 2156 as the product of its prime factors. [2]

b) Find the highest common factor of 2156 and 385. [2]

c) Write down an irrational number that lies between 3 and 4. [1]

d) Yumi says that to divide a number by a fraction you can multiply the number by the fraction's reciprocal. Explain why this is true. [2]

2
a) Which of these improper fractions has the largest value? [2]

$$\frac{19}{4} \qquad \frac{25}{6} \qquad \frac{45}{11}$$

Extended

b) Convert the recurring decimal $0.0\dot{2}\dot{5}$ to a fraction in its simplest form. [3]

c) A real estate firm takes 1.2% of the sales value of a house as its commission. What is the commission on a house that sells for $475 000? [2]

d) A painting, bought six years ago for $10 500, sold for $10 899. What is the percentage increase in the painting's selling price? [3]

e) A new road has opened which cuts the journey time between two towns. The journey used to take 2 hours and 30 minutes but now takes 18% less time. How long is the new journey time? [2]

Extended

f) Katya invests €25 000 for five years at 1.2% compound interest. How much will Katya's investment be worth at the end of the five years? [3]

g) Zack's market stall sees an increase of 38% in weekly sales during the summer season. If he takes $1766.40 during a summer week, how much did he take during a week before the summer? [3]

3
a) Alice is organising a trip to an international netball match. A coach, for 65 people, costs $350 for the trip. There are 384 people interested in going on the trip. Alice wants to make $75 for her netball club funds. How much should Alice charge each person for the trip? [3]

b) Lucas has three sacks of potatoes weighing $4\frac{2}{7}$ kg, $5\frac{1}{6}$ kg and $3\frac{2}{3}$ kg. He gives $12\frac{1}{2}$ kg of potatoes to Rana. How many kilograms does Lucas have left? [3]

c) The width of a shop is 11 metres. Jim's shelving units are each $1\frac{7}{8}$ m wide. How many complete units will fit into the available width? [2]

How much will Jim need to cut off a shelving unit to fill the last space? [1]

4
a) A long-haul flight from Los Angeles to Copenhagen flies over Ilulissat in Greenland. One March morning the temperature in Ilulissat was −21 °C. On board the aircraft the cabin temperature was 24 °C. What was the difference in temperature at this point? [2]

b) Hannah parked her car on level −5 of the carpark and took the lift to her office 18 floors above. Later, she walked down three floors to have lunch in the company dining suite. After lunch she had a meeting with a business partner and she took the lift from the dining suite up eight floors. On what floor was the meeting? [2]

5 a) Put these in order starting with the smallest first. [2]

17^2 $\sqrt{300}$ $(-16.8)^2$ $\sqrt{280}$

b) Find two consecutive integers between which $\sqrt[3]{-400}$ lies. [2]

Extended

c) In 2012 the population of Ganges river dolphin was estimated to be 1800. It was also estimated that the population was declining by 7.5% per year. If no conservation work was done to halt this decline, how many years would it take for the population to drop below 900? [3]

6 a) n is an odd number and $30 < n \leq 40$. Given that n is not a multiple of 3, list all possible values of n. [2]

b) Look at this Venn diagram. List the elements of:

i) $A \cap B$ ii) $A \cup B$ [1, 1]

c) Find $n(A \cap B)$ [1]

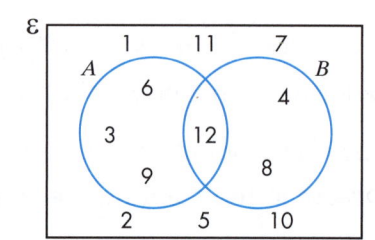

Extended

d) List the elements of: i) $(A \cup B)'$ ii) $A \cap B'$ [2, 2]

e) $A = $ {prime numbers between 80 and 100}. How many proper subsets of A are there? [2]

7 a) A certain shade of green paint requires blue and yellow to be mixed in the ratio 3 : 5. A painter has 8 litres of yellow paint and an unlimited quantity of blue paint. How many litres of green paint can he make? [2]

Extended

b) A house, previously valued at $268 800, has decreased in value in the ratio 11 : 12. What is the current value of the house? [2]

c) To win the Triathlon gold medal at the Rio Olympics, Alistair Brownlee swam 1.5 km in 17 minutes, cycled 40 km in 55 minutes and ran 10 km in 31 minutes. What was his average speed, in km/h? [4]

d) A steam train needed approximately 150 kg of coal per km travelled.

   i) How much coal, in tonnes, did it need to travel from London to Edinburgh (632 km)? [2]

   ii) How many kilometres per tonne of coal does the train travel (to 1 d.p.)? [2]

   iii) Would a similar train with 220 tonnes of coal complete the journey from New Delhi to Hyderabad (1660 km)? [2]

IGCSE Mathematics Revision Guide

e) A square-shaped children's swimming pool is being filled with water through a hosepipe. The swimming pool has a side length of 210 cm and is 100 cm deep. The hosepipe has a flow rate of 20 litres/minute. How deep is the water after 60 minutes, in cm (to 3 s.f.)? [3]

How many minutes will it take to fill the swimming pool? [2]

f) Pete's Petshop sells 25 kg of cat food for $16. Kirsty's Kennels sells the same cat food in 20 kg sacks for $12.60 each. Which of the two shops has the best deal? [3]

g) An online agency promotes clients' advertisements on websites. Their charges depend on the popuarity of the websites. A client has paid a fixed fee for an advert to be displayed. On website A the advert is displayed 50 000 times at a price of 0.5p every time their page is loaded. The charge for website B is 1.25p every time their page is loaded. How many times could the client have their advert displayed on website B for the price to be the same as website A? [3]

h) A toy box has 100 wooden blocks. Each block measures 2.5 cm × 2.5 cm × 10 cm. The blocks have a total mass of 4.7 kg (2 s.f.). Kai says that the blocks are made from the wood of a holly tree (density 0.75 g/cm^3), but Lea says that they are made of apple wood (density 0.65 g/cm^3). Who is correct? [1, 2]

i) A wooden packing crate measures 1.5 m by 75 cm by 1 m. On which face should it be stood to exert least pressure? [1]

8 a) Round 527.815 to 2 s.f., 2 d.p., 4 s.f. and the nearest integer. [1, 1, 1, 1]

Extended

b) The contents of a sack of sand weigh 15 kg, to the nearest 100 g. What are the lower and upper bounds for the amount of sand in the sack? [2]

c) A soup ladle holds a 475 ml serving of soup, measured to the nearest 25 ml. A cooking pot holds 25 litres, measured to the nearest 100 ml, of soup. What are the minimum and maximum number of servings in the cooking pot, to the nearest serving? [3]

9 a) Write the numbers 25 400 000 and 0.000 254 in standard form. [2, 2]

b) The average distance between the planets Mercury and Venus is 5.029×10^7 km and between Mercury and Mars it is 170 030 000 km. Which is further away from Mercury? [1]

By how many kilometres? [1]

Extended

10 a) A jar contains 1.5 litres of maple syrup. The density of maple syrup is 1.37 g/cm³. The glass jar has a mass of 750 g. What is the mass, in kilograms, of the jar of syrup? [2]

b) Ivan travelled from Paris to Ho Chi Minh City. When he left Paris the exchange rate was €1 = 24 850 Vietnamese dong. When Ivan returned to Paris the rate was €1 = 25 650 Vietnamese dong. Ivan left Paris with €2500. When he returned to Paris Ivan exchanged his spare dong and received €315. How many dong had Ivan spent? [2]

c) Work out the answer to this calculation, writing down all the figures on your calculator display. [2]

$$\sqrt[3]{\frac{22.8^2 + 9.45^3}{22.8 \times 9.45}}$$

Algebraic representation and formulae

The language of algebra

Whenever you make a calculation with numbers, for example, working out your change after buying something in a shop, the result is always another number. With algebra you use letters to represent values which, at the time are unknown. Later, you can either substitute a value for the letter when it becomes known or use other information to work out what the unknown value is.

Make sure that you know and understand this important algebra vocabulary.

- **Variable:** the letters that are used to represent the unknown, varying, value.
- **Expression:** the combination of letters and numbers that are used to describe or explain a problem. $3x^2 - 7t + 5$
- **Equation:** when an expression can be equated to something it has an equals sign added to it. This means that the equation can be solved to find a value for the variable. $3x + 8 = 14$
- **Formula:** these are equations that may contain more than one variable and also be used as rules to solve particular problems such as finding the volume of a sphere. $V = \frac{4}{3}\pi r^3$
- **Term:** these are the separate parts of an expression, equation or formula. The expression, $3x^2 - 7t + 5$, contains two variables: x and t, and three terms: $3x^2$, $-7t$ and $+5$

You also need to be able to express problems with unknown quantities algebraically.

> **Revision tip**
>
> Remember that $6m$ means 6 times m, you do not need to write the multiplication sign between the number and letter.

Example

1 a) The triangle has three sides with lengths as shown in the diagram.

The perimeter, P, of the triangle is the sum of all three sides.

$P = (x + 3) + (10x - 7) + 6x$
$P = 17x - 4$

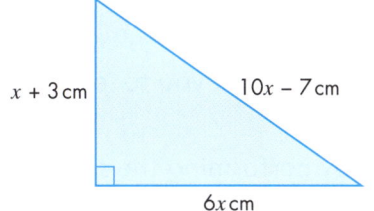

$x + 3$ cm $10x - 7$ cm

$6x$ cm

b) The area, A, of a triangle is found using the formula: $\frac{1}{2} \times$ base length $\times$ vertical height, or $A = \frac{1}{2}bh$.

$A = \frac{1}{2} \times 6x(x + 3)$
$A = 3x(x + 3)$

2 Alice, Emily and Hannah receive pocket money every week. Alice receives $3.50 more than Hannah, who receives $2 more than Emily.

If Emily receives x, Hannah receives $x + 2$, and Alice receives $(x + 2) + 3.5$, or $x + 5.5$

If the total pocket money they receive is y, you can write:

$y = A + E + H$
$y = (x + 5.5) + (x + 2) + x$
$y = 3x + 7.5$

Substitution into formulae

The terms used in a formula do not change, because they describe the solution to a problem, but the values of the variables within the formula can change.

In the example on p41 for finding the area of a triangle: $A = \frac{1}{2}bh$, the formula always works. Finding the area of any triangle depends only on substituting values for the base length, b, and the vertical height, h.

If the value of x had been 2, the base length would be $6 \times 2 = 12$, and the vertical height $2 + 3 = 5$

Substituting these into the formula: $A = \frac{1}{2}bh$, gives $A = \frac{1}{2} \times 12 \times 5$, so $A = 30$

If the weekly amount of pocket money received by Emily, page 41, was $12.50 you could work out the value of the money given by the girls' parents as:

$$y = 3 \times 12.5 + 7.5$$
$$y = 45$$

Hence, the value of the money given by the girls' parents was $45

Often the word 'evaluate' is used instead of 'find the value of'.

Rearranging formulae

The **subject** of a formula is the variable in the equation or formula that is a single letter on one side of the equals sign, usually on the left-hand side.

In $A = \frac{1}{2}bh$, A is the subject. In $V = \frac{4}{3}\pi r^3$, V is the subject.

If another variable is required to be the subject, you need to **rearrange** the equation or formula so that the required variable stands by itself on the left-hand side.

For example, if in $A = \frac{1}{2}bh$, you are required to make b the subject of the equation, you can rearrange it as:

$$b = \frac{2A}{h}$$

Another way of asking the same thing is to ask you to 'express b in terms of A and h'.

Remember to balance the equation by performing the same operation on both sides of the equation, as shown in the example below.

> **Revision tip**
>
> Remember inverse operations, and the rule 'whatever you do to one side of the equation you must do to the other side'.

Example

Make x the subject in these equations.

a) $3x + 7 = t$

   $3x = t - 7$ Subtract 7 from both sides.

   $x = \dfrac{t - 7}{3}$ Divide both sides by 3.

b) $A = 2y + 3x^2$

   $A - 2y = 3x^2$ Subtract $2y$ from both sides.

   $\dfrac{A - 2y}{3} = x^2$ Divide both sides by 3.

   $\sqrt{\dfrac{A - 2y}{3}} = x$ Take the square root of both sides.

Extended

More complicated formulae

To find the value of a variable, you need to make sure it is the subject of the formula. This might mean rearranging the formula first, which may take a number of steps.

Example

$w = \sqrt{2c + 3d}$

Make d the subject and find the value of d when $w = 8$ and $c = 20$.

Square both sides: $\quad\quad\quad w^2 = 2c + 3d$

Subtract $2c$ from both sides. $\quad w^2 - 2c = 3d$

Divide both sides by 3. $\quad\quad \dfrac{w^2 - 2c}{3} = d$ so $d = \dfrac{w^2 - 2c}{3}$

Then $d = \dfrac{8^2 - 2 \times 20}{3} \rightarrow d = \dfrac{64 - 40}{3} \rightarrow d = \dfrac{24}{3}$ so $d = 8$

Quick test

1. Write expressions for the total cost of:
   a) 3 cinema tickets at $7.50 each
   b) x cinema tickets at $7.50 each
   c) x cinema tickets at y each.

2. Katya's car is x years older than Daria's. Write an expression for the age of Daria's car.

3. Gustav has b motorbikes in his collection.
   a) How many wheels are there in total on Gustav's motorbikes?
   b) Modifications are made to five of the bikes so that they have a spare wheel mounted on the back. What is the total number of wheels now?
   c) Gustav acquires y tricycles that have three wheels. What is the total number of wheels now?

4. Find the value of $\sqrt{m^2 + n^2}$ when: a) $m = 5$ and $n = 12$ b) $m = 7$ and $n = 10$ (to 2d.p.).

5. The cost, C ($), of hiring a bouncy castle from a company who organises parties is:

   $$C = 4H + 1.25D + 25$$

   where H is the number of hours the bouncy castle is required and D (km) is the distance that the company needs to travel to deliver the bouncy castle.
   a) Natalie wants to hire the castle for 10 hours and lives 20 km away from the company. How much does she pay?
   b) Rearrange the formula to make H the subject. Clive paid $168.75 and lives 35 km from the company. For how many hours did he hire a bouncy castle?
   c) Rearrange the original formula to make D the subject. Rita paid $159.90 to hire a bouncy castle for 18 hours. How far does she live from the company?

Algebraic manipulation 1

Simplifying expressions

To **simplify** an expression is to make it shorter and easier to read. Terms in an expression can be combined and collected together.

To simplify by multiplying, first multiply any numbers together and then multiply the letters together. By convention, you write numbers first and then the letters in alphabetical order.

$6 \times m = 6m$ $\qquad\qquad 3m \times 2t = 6mt$

$4w \times 5w = 20w^2$ $\qquad 3vz \times 2tz = 6tvz^2$

$2e \times -3ef = -6e^2f$

The number in front of the letter representing the variable is the **coefficient**.

Collecting like terms

Like terms are those that have the same combination of variables. Collecting like terms involves adding or subtracting the coefficients.

$2b + 3c + 6b - c = 2b + 6b + 3c - c = 8b + 2c$

$3ab + 7ab - 3bc - 2ba = 3ab + 7ab - 2ab - 3bc = 8ab - 3bc$ (note that $ab = ba$)

$5x^2 - 3x^2 = 2x^2$

Expanding brackets

To **expand** brackets you multiply them out. The resulting expression does not contain any brackets. The term that is outside the brackets must be multiplied by every term that is inside the brackets. The resulting expression does not contain any brackets.

$3(t - 5) = 3t - 15$

$v(3 + 5d) = 3v + 5vd$

$3v(v + 2t) = 3v^2 + 6tv$

$2w^2(1 - 5h) = 2w^2 - 10hw^2$

$-b(f + 3g) = -bf - 3bg$

$2h^2(h - 3k) = 2h^3 - 6h^2k$

$-4m(3n - 2m) = -12mn + 8m^2$

Expanding and simplifying

When you expand brackets, some of the terms in the resulting expression can usually be combined. Always simplify expressions as much as you can when you multiply out brackets.

$3(t - 5) + 5(t + 2) = 3t - 15 + 5t + 10 = 3t + 5t - 15 + 10 = 8t - 5$

$6(p + 5q) - 4(p + q) = 6p + 30q - 4p - 4q = 6p - 4p + 30q - 4q = 2p - 26q$

$2m(3 + 6n) + 4n(2 - m) = 6m + 12mn + 8n - 4mn = 6m + 12mn - 4mn + 8m = 6m + 8mn + 8n$

$3f(2f - 3g) - f(-f + 6g) = 6f^2 - 9fg + f^2 - 6fg = 6f^2 + f^2 - 9fg - 6fg = 7f^2 - 15fg$

Factorisation

To factorise an expression, do the opposite of expansion. Factorising an expression will introduce brackets.

To factorise, you need to look for the lowest common multiple of all the terms. This is the term that will go outside the brackets. The terms that go inside the brackets are the results of removing the common factors.

In the expression $10x + 5y$, 5 is a common factor: $5(2x) = 10x$ and $5(y) = 5y$.

So, 5 goes outside the brackets and $2x$ and y go inside: $5(2x + y)$

You can always check that your factorising works by expanding your answer: you should get the expression that you started with.

In $4h^2 - 12h$, $4h$ is common, so: $4h(h - 3)$

In $8m^2n + 20mn^2$, $4mn$ is common, so: $4mn(2m + 5n)$

> **Revision tip**
>
> Factorising and expanding are the reverse of each other. This means that you can check your answer by carrying out the reverse operation to make sure you get back to the question.

> **Revision tip**
>
> Be particularly careful with the signs when multiplying elements in the brackets.

Multiplying two brackets

In a **quadratic expression**, the highest power of the variables is 2.

Expressions such as $x^2 - 9$, $x^2 + 5x + 4$ and $6x^2 - 5x - 4$, are all quadratic expressions.

These expressions may be the result of expanding two brackets, each with two terms in them. For example, $(x + 3)(x - 3)$, $(x + 1)(x + 4)$ and $(2x + 1)(3x - 4)$ expand to give the expressions above.

Remember that $(x + 2)^2$ is actually $(x + 2)(x + 2)$.

Expanding these brackets is called quadratic expansion. Each term in the left bracket must be multiplied by each term in the right bracket.

There are a number of methods that can be used to expand a set of two brackets, the three main ones being expand, FOIL (First, Outer, Inner, Last) and box or grid method.

Example

Expand $(3x + 2)(4x - 3)$.

a) Expand:

$$(3x + 2)(4x - 3) = 3x(4x - 3) + 2(4x - 3)$$
$$= 12x^2 - 9x + 8x - 6$$
$$= 12x^2 - x - 6$$

b) FOIL:

$(3x + 2)(4x - 3)$

$12x^2 - 9x + 8x - 6$
$= 12x^2 - x - 6$

c) Box or grid method:

	$3x$	$+2$
$4x$	$12x^2$	$8x$
-3	$-9x$	-6

$12x^2 - 9x + 8x - 6$
$= 12x^2 - x - 6$

Algebra 45

Extended

Multiplying three brackets

If an expression such as $(x + 1)(x + 2)(x - 1)$ needs expanding then you should multiply one pair of brackets together and then multiply the result by the remaining bracket.

$$(x + 1)(x + 2)(x - 1) = (x + 1)[(x + 2)(x - 1)]$$
$$= (x + 1)[x^2 + x - 2]$$
$$= x^3 + 2x^2 - x - 2$$

Quick test

1. Simplify these expressions.
   a) $4 \times v$
   b) $4w \times 2p$
   c) $4q \times 5q$
   d) $2fg \times 2gh$
   e) $2d \times -3de$

2. Simplify these expressions by collecting the like terms.
   a) $3p + 3r + 6p - r$
   b) $6cd - 5cd - 3tv + 3dc$
   c) $5w^2 + 2w^2$
   d) $2a^2 - 6g^3 + 7g^3 - 3a^2$

3. Expand and simplify these expressions.
   a) $4(h + 2)$
   b) $v(6 - 4t)$
   c) $q(3 - 2r) - 5(v + q)$
   d) $2g(3h + 4g) - g(-g - 2h)$

4. Factorise fully these expressions.
   a) $16p - 12q$
   b) $12s + 9st$
   c) $20g^2h - 14fg$
   d) $9cd^2 - 15c^2d^3e$
   e) $16mn^3 + 8mn$

5. Expand and simplify these expressions.
   a) $(x + 6)(x - 2)$
   b) $(2w - 4)(3w + 2)$
   c) $(p + 3)^2$
   d) $(2 + 3m)(6 - 2m)$
   e) $(q - 2)(q + 2)$

Extended

6. Expand $(x - 2)^3$

Algebraic manipulation 2

Extended

Quadratic factorisation

Factorising is the opposite of expansion. It means putting brackets back into an expression. With quadratic factorisation this will involve finding two sets of brackets.

Consider $(x + 1)(x + 4)$, which expands to $x^2 + 5x + 4$. Notice that the factors $1 + 4$ give the middle term and 1×4 gives the final term.

In $x^2 + 7x + 10$, the product of the factors 2 and 5 is 10 (the final term) and the sum of the factors gives the middle term.
So, $x^2 + 7x + 10 = (x + 2)(x + 5)$.
$x^2 + 6x - 16 = (x + 8)(x - 2)$
$x^2 - 9x + 14 = (x - 2)(x - 7)$
$x^2 - 3x - 18 = (x - 6)(x + 3)$

> **Revision tip**
>
> First decide on the signs that will be needed in the brackets, then look at the numbers.
> These are the key aspects when looking to factorise the quadratic expression.

A quadratic expression is of the form $ax^2 + bx + c$, where a, b and c are constants with a not equal to 0.

The method outlined above can be adjusted slightly to take into account the coefficient of x^2 when a is not $= 1$.

Consider $2x^2 + 9x + 10$.

$2x^2$ can only be factorised as $2x \times x$ and the signs in the brackets must be positive.

The brackets must be $(2x + ?)(x + ?)$

+10 has two factor pairs: 1 and 10, or 2 and 5, so you need find the combination that will give a middle term of $9x$.

$(2x + 1)(x + 10) = 2x^2 + 20x + x + 10 = 2x^2 + 21x + 10$ ✘

$(2x + 10)(x + 1) = 2x^2 + 2x + 10x + 10 = 2x^2 + 12x + 10$ ✘

$(2x + 2)(x + 5) = 2x^2 + 10x + 2x + 10 = 2x^2 + 12x + 10$ ✘

$(2x + 5)(x + 2) = 2x^2 + 4x + 5x + 10 = 2x^2 + 9x + 10$ ✓

Factorised, $2x^2 + 9x + 10 = (2x + 5)(x + 2)$

In the same way:

$3x^2 - 7x - 6 = (3x + 2)(x - 3)$
$2x^2 - 6x - 8 = (2x + 2)(x - 4)$
$4x^2 + 16x + 15 = (2x + 3)(2x + 5)$

The difference of two squares

Quadratic expressions such as **a)** $x^2 - 1$; **b)** $x^2 - 16$; **c)** $m^2 - 64$; **d)** $169 - w^2$; **e)** $9p^2 - 49$, have a distinctive form and are called the **difference of two squares**. There are two terms, each being a perfect square, and one term is subtracted from the other.

The factorisation is: $a^2 - b^2 = (a - b)(a + b)$.

When $(a - b)(a + b)$ is expanded, $a^2 + ab - ab - b^2$, the middle terms always cancel out.

The expressions above will factorise to: **a)** $(x - 1)(x + 1)$; **b)** $(x - 4)(x + 4)$; **c)** $(m - 8)(m + 8)$; **d)** $(13 - w)(13 + w)$; **e)** $(3p - 7)(3p + 7)$

Algebraic fractions

Working to simplify an **algebraic fraction** requires you to use all the skills presented so far: expanding, factorising, simplifying, rearranging. It is useful to recall the operations and methods you use when working with simple numeric fractions.

When asked to simplify an algebraic fraction you are required to rewrite the expression as a single fraction or even a single **linear** expression.

> **Example**
> a) $\dfrac{x+2}{3} - \dfrac{x-1}{4}$
>
> $= \dfrac{4(x+2) - 3(x-1)}{12}$ The common denominator is 12 so multiply the numerator of the first fraction by 4 and the numerator of the second fraction by 3.
>
> $= \dfrac{4x + 8 - 3x + 3}{12}$ Expand and simplify the numerator.
>
> $= \dfrac{x + 11}{12}$
>
> b) $\dfrac{x^2 - x - 12}{2x^2 + 2x - 12}$
>
> $= \dfrac{(x-4)(x+3)}{(2x-4)(x+3)}$ Factorise the numerator and the denominator.
>
> $= \dfrac{x-4}{2x-4}$ Cancel the common factor.

Quick test

Extended

1. Factorise these expressions.
   a) $m^2 - 2m - 24$ b) $6 + y - y^2$ c) $v^2 - 4v + 3$ d) $w^2 + 8w + 16$ e) $p^2 - 3p - 18$

2. Factorise these expressions.
   a) $p^2 - 49$ b) $q^2 - 121$ c) $w^2 - 1$ d) $225 - d^2$ e) $e^2 - f^2$

3. Simplify these expressions.
   a) $\dfrac{x}{5} + \dfrac{2x}{3}$ b) $\dfrac{x+2}{5} - \dfrac{x-3}{2}$ c) $\dfrac{x}{9} \times \dfrac{3}{2x}$ d) $\dfrac{4x}{15} \div \dfrac{5xy}{6}$ e) $\dfrac{x^2 - 4}{2x^2 - 7x + 6}$

Solutions of equations and inequalities 1

Solving linear equations

An equation is formed when one expression is made equal to a number or another expression. The solution to the equation is the value of the variable that makes the equation true.

The key to solving an equation successfully is to remember to carry out the same operation to both sides of the equation. This is sometimes called the balance method.

Example

a) $5m + 8 = 23$

Subtract 8 from both sides: $5m + 8 - 8 = 23 - 8 \rightarrow 5m = 15$

Divide both sides by 5: $\dfrac{5m}{5} = \dfrac{15}{5} \rightarrow m = 3$

b) $\dfrac{5w - 6}{7} = 2$

Multiply both sides by 7: $\dfrac{7(5w - 6)}{7} = 7 \times 2 \rightarrow 5w - 6 = 14$

Add 6 to both sides: $5w - 6 + 6 = 14 + 6 \rightarrow 5w = 20$

Divide both sides by 5: $\dfrac{5w}{5} = \dfrac{20}{5} \rightarrow w = 4$

Whenever brackets appear in an equation, you should expand them first.

Example

a) $6(2x - 7) = -18$

Expand the brackets: $12x - 42 = -18$

Add 42 to both sides: $12x - 42 + 42 = -18 + 42 \rightarrow 12x = 24$

Divide both sides by 12: $\dfrac{12x}{12} = \dfrac{24}{12} \rightarrow x = 2$

A second method would be to divide both sides by the factor outside the brackets:

Divide both sides by 6: $\dfrac{6(2x - 7)}{6} = -\dfrac{18}{6} \rightarrow 2x - 7 = -3$

Add 7 to both sides: $2x - 7 + 7 = -3 + 7 \rightarrow 2x = 4$

Divide both sides by 2: $\dfrac{2x}{2} = \dfrac{4}{2} \rightarrow x = 2$

> **Revision tip**
>
> It is better to rearrange before simplifying. Trying to do them at the same time can mean that mistakes are made.

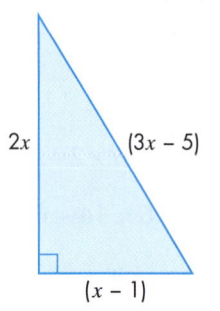

Sometimes the equation has the variable on both sides of the equation. In these cases you need to collect together all the terms that contain the variable on one side of the equals sign.

> **Example**
> a) $4(q + 3) = 7q - 3$
> Expand the brackets: $\quad 4q + 12 = 7q - 3$
> Subtract $7q$ from both sides: $4q - 7q + 12 = 7q - 7q - 3 \to -3q + 12 = -3$
> Subtract 12 from both sides: $-3q + 12 - 12 = -3 - 12 \to -3q = -15$
> Divide both sides by -3: $\quad \dfrac{-3q}{-3} = \dfrac{-15}{-3} \to q = 5$

Setting up equations

Equations can be written to represent a situation.

> **Example**
> a) This triangle has a perimeter of 30 cm. What are the lengths of the three sides?
> The perimeter is: $2x + (x - 1) + (3x - 5) = 30$
> $6x - 6 = 30$
> $6x = 36$
> $x = 6$
>
> Sides labelled: $2x$, $(3x - 5)$, $(x - 1)$
>
> Substitute $x = 6$ into the expressions to find the side lengths are: 5 cm, 12 cm, 13 cm
>
> b) A taxi company charges an amount per mile with a fixed call-out charge of $3. Jabir is charged $5.16 for a 12-mile journey. What is the charge per mile?
> Let m be the charge per mile: $\quad 12m + 3 = 5.16$
> Subtract 3 from both sides: $\quad 12m + 3 - 3 = 5.16 - 3 \to 12m = 2.16$
> Divide both sides by 12: $\quad \dfrac{12m}{12} = \dfrac{2.16}{12} \to m = \$0.18 \text{ (or 18p)}$

Extended

Solving inequalities

You solve linear inequalities in the same way as you solve linear equations. In 'Ordering and set notation' (page 21) you were introduced to the inequality symbols: $<, \leq, >, \geq$.

When solving an inequality, never replace the inequality symbol with an equals sign.

For example, suppose you need to solve $4x + 3 < 15$.

Subtract 3 from both sides: $\quad 4x + 3 - 3 < 15 - 3 \to 4x < 12$

Divide both sides by 4: $\quad \dfrac{4x}{4} < \dfrac{12}{4} \to x < 3$

This means that x can take any value less than 3.

There is one situation that needs great care. If you divide or multiply by a negative value you need to reverse the inequality sign: $<$ and $\leq$ become $>$ and $\geq$, and similarly $>$ and $\geq$ become $<$ and $\leq$.

> **Revision tip**
> If possible avoid multiplying and dividing by a negative number, so that the inequality sign remains the same throughout the solution.

Example

Solve $7 - \frac{8x}{3} \leq 3$

Subtract 7 from both sides: $\quad 7 - 7 - \frac{8x}{3} \leq 3 - 7 \rightarrow -\frac{8x}{3} \leq -4$

Multiply both sides by -3: $\quad -3\left(-\frac{8x}{3}\right) \geq -3 \times 3 \rightarrow 8x \geq 12 \qquad$ [Note the reversal: $\leq$ to $\geq$]

Divide both sides by 8: $\quad x \geq 1.5$

Representing linear inequalities on a number line

Solutions to a linear equation can be shown on a number line using these symbols:

A solid circle means the value is included.

An open circle means the value is not included.

Using these symbols:

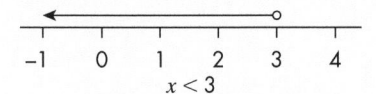

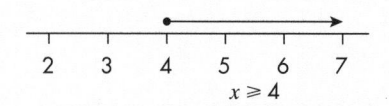

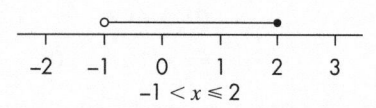

$x < 3$
$x \geq 4$
$-1 < x \leq 2$

For example, a rectangular badge has side lengths that are x cm and $(x - 3)$ cm and its perimeter cannot be more than 20 cm. Represent the possible solutions for x on a number line.

First, derive the inequality:

$x + x + (x - 3) + (x - 3) \leq 20$

$4x - 6 \leq 20$

Solve the inequality:

$4x \leq 26$

$x \leq 6.5$

Also, since $(x - 3)$ is the length of one side, $x - 3 > 0$, so $x > 3$

Putting these two inequalities together: $3 < x \leq 6$

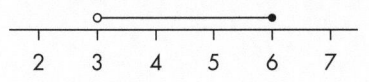

Quick test

1. Solve these equations.

   a) $3w + 2 = 14$ b) $2(3t + 4) = 20$ c) $12v - 5 = 4 - 6v$ d) $5(2d + 6) = 3(7 + 2d) + 1$

2. Advice for cooking a turkey is 90 minutes plus m minutes per kilogram meaning that a 15 kg turkey is cooked for 6 hours 30 minutes. How long would it take to cook an 18 kg turkey?

Extended

3. Solve these inequalities.

   a) $w + 6 > 5$ b) $3v - 4 \leq 2$ c) $2(4t + 1) \geq -6$ d) $6 - \frac{3x}{4} > 3$

4. Solve these inequalities and represent their solutions on number lines:

   a) $5x - 6 > 9$ b) $2(3x - 4) \leq 4$

Algebra

Simultaneous equations

Solving **simultaneous equations** means finding a solution that satisfies both equations at the same time. There are two methods: **elimination** and **substitution**. The method to use depends on the coefficients of the variables and the way that the equations are written – and your own confidence with them.

$$6x + 2y = 14 \quad (1)$$
$$3x - 5y = 10 \quad (2)$$

Elimination method

First, you need a coefficient for one of the variables to be the same in both equations. Multiply (2) throughout by 2 giving a coefficient for x of 6, resulting in (3).

$$6x + 2y = 14 \quad (1)$$
$$6x - 10y = 20 \quad (3)$$

If the original equations have a coefficient that is the same for one of the variables, this step is not required. In some questions you might need to multiply both equations to obtain equations that have the same coefficient.

In this case, the signs of the x terms are the same (both positive) and so subtract to **eliminate** the x terms. If the signs had been different to each other, as in the terms for y, the equations would need to be added together.

$$(1) - (3) = 12y = -6 \rightarrow y = -\frac{1}{2}$$

Now substitute $y = -\frac{1}{2}$ into one of the original equations to find x.

Substitute into (1): $6x + 2\left(-\frac{1}{2}\right) = 14 \rightarrow 6x = 15 \rightarrow x = 2.5$

> **Revision tip**
>
> When the signs of the variable being eliminated are the same subtract the equations; when they are different, add the equations.

Substitution method

Since (1) contains even numbers for each term you can divide throughout by 2 to get a coefficient of 1 for the y term: $3x + y = 7$ (4)

Rearrange (4) to make y the subject: $y = 7 - 3x$ (4), and substitute this into (2).

$$3x - 5(7 - 3x) = 10 \rightarrow 3x - 35 + 15x = 10 \rightarrow 18x = 45 \rightarrow x = 2.5$$

As before, substitute $x = 2.5$ into one of the original equations to find $y = -\frac{1}{2}$

Problem solving using simultaneous equations

Sometimes you will only be given the outline of a problem. This means that you will need to derive (create) the equations from the information that has been given, and then solve them.

For example: Jamilla buys 4 loaves and 3 cakes for $8.55; Svetlana buys 5 loaves and 2 cakes for $9.20 Find the cost of 2 loaves and 5 cakes.

Jamilla:	$4L + 3C = 8.55$ (multiply by 2)	$8L + 6C = 17.10$
Svetlana:	$5L + 2C = 9.20$ (multiply by 3)	$15L + 6C = 27.60 -$
		$7L = 10.50$
		$L = 1.50$

If a loaf costs $1.50, then $4(1.5) + 3C = 8.55 \rightarrow 3C = 2.55 \rightarrow C = 0.85$, a cake is $0.85; 2 loaves and 5 cakes will be: $2(1.5) + 5(0.85) = \$7.25$

Extended

More complex simultaneous equations

When a pair of simultaneous equations includes one which is a quadratic, there could be more than one pair of solutions. To solve these types of simultaneous equations the easiest method is substitution.

Example

$$x + y = 1 \quad (1)$$
$$x^2 - xy = 6 \quad (2)$$

Rearrange (1): $y = 1 - x$

and substitute this into (2): $x^2 - x(1 - x) = 6$

Solve for x: $x^2 - x + x^2 = 6$

$2x^2 - x - 6 = 0$

$(2x + 3)(x - 2) = 0$

$2x + 3 = 0 \rightarrow x = -1.5$

$(x - 2) = 0 \rightarrow x = 2$

Since there are two solutions for x there will be two solutions for y.

Substitute both solutions into (1)

If $x = -1.5 \rightarrow -1.5 + y = 1 \rightarrow y = 2.5$

If $x = 2 \rightarrow 2 + y = 1 \rightarrow y = -1$

The solutions are: $x = -1.5$, $y = 2.5$ and $x = 2$, $y = -1$

Quick test

1. Solve these simultaneous equations.

    a) $5x + 2y = 4$
       $3x - 2y = 0$

    b) $3x + 4y = 1$
       $5x - 2y = -20$

    c) $4x + y = 11$
       $2x + 3y = 13$

    d) $3x - 8y = 16$
       $2x + 6y = 5$

2. The Patel family (2 adults and 3 children) went to the cinema. Their tickets cost $39.00
   They met the Jaing family (1 adult and 2 children) who had paid $22.50
   How much would tickets for 3 adults and 2 children cost?

Extended

3. Solve these simultaneous equations: $x^2 + y^2 = 5$ and $x + y = 3$

Solutions of equations and inequalities 2

Extended

Solving quadratic equations by factorisation

In the previous section you practised factorising a quadratic expression. You need to use the same skills to begin to solve a quadratic equation. Make sure that the equation is equal to zero before factorising – this may mean that you need to rearrange the equation first. For example:

$x^2 + 11x + 18 = 0$ factorises to $(x + 2)(x + 9) = 0$

This can only be true when either $(x + 2) = 0$, or when $(x + 9) = 0$

Meaning that: $x + 2 = 0 \rightarrow x = -2$ or

$x + 9 = 0 \rightarrow x = -9$

$6x^2 - 9x + 1 = 4 - 2x$ rearranges to $6x^2 - 7x - 3 = 0$

and then factorises to: $(3x + 1)(2x - 3) = 0$

This can only be true when either $(3x + 1) = 0$, or when $(2x - 3) = 0$

Meaning that: $3x + 1 = 0 \rightarrow x = -\frac{1}{3}$ or

$2x - 3 = 0 \rightarrow x = 1.5$

> **Revision tip**
>
> Always rearrange a quadratic to the form $ax^2 + bx + c = 0$ before trying to factorise or use the quadratic formula.

Solving quadratic equations by the quadratic formula

When a quadratic equation will not factorise because it does not have simple integer factors you need to use the **quadratic formula**: $x = \dfrac{-b \pm \sqrt{b^2 - 4ac}}{2a}$

a and b are the coefficients of x^2 and x, and c is the constant term.

To solve $x^2 + 14x + 12 = 0$, $a = 1$, $b = 14$, $c = 12$

Use $x = \dfrac{-b \pm \sqrt{b^2 - 4ac}}{2a}$: $x = \dfrac{-14 \pm \sqrt{14^2 - 4 \times 1 \times 12}}{2 \times 1} \rightarrow x = \dfrac{-14 \pm \sqrt{148}}{2}$

So $x = -0.92$ or -13.08 (2 d.p.)

Solve $5x^2 + 3x - 7 = 0$

$a = 5$, $b = 3$, $c = -7$

Use $x = \dfrac{-b \pm \sqrt{b^2 - 4ac}}{2a} \rightarrow x = \dfrac{-3 \pm \sqrt{3^2 - 4 \times 5 \times (-7)}}{2 \times 5} \rightarrow x = \dfrac{-3 \pm \sqrt{149}}{10}$

$x = 0.92$ or -1.52 (2 d.p.)

Solving quadratic equations by completing the square

This is an alternative method for solving quadratic equations.

The general quadratic expression $x^2 + px + q$ can be written the form $(x + a)^2 + b$ using $\left(x + \dfrac{p}{2}\right)^2 - \left(\dfrac{p}{2}\right)^2 + q$

$x^2 + 14x \rightarrow (x + 7)^2 - 49$

$x^2 + 14x + 12 \rightarrow (x + 7)^2 - 49 + 12 \rightarrow (x + 7)^2 - 37$

Quick test

Extended

1. Factorise and then solve these quadratic equations.
   a) $x^2 + 11x + 28 = 0$ b) $z^2 - 2z - 15 = 0$ c) $2s^2 - 2s - 12 = 0$ d) $9q^2 - 16 = 0$
2. Solve these quadratic equations, using the quadratic formula, giving your answers to 2 d.p.
   a) $x^2 + 10x - 4 = 0$ b) $x^2 - 8x + 6 = 0$ c) $2x^2 - 9x = 3x + 2$ d) $5x^2 + 18x = 0$
3. Solve each equation by completing the square. Give your answers in surd form.
   a) $x^2 + 10x = 0$ b) $x^2 + 10x - 4 = 0$ c) $x^2 - 4x - 6 = 0$ d) $x^2 - 15x + 2 = 0$
4. Solve each equation by completing the square. Give your answers to 2 d.p.
   a) $x^2 + 12x = 0$ b) $x^2 + 12x - 9 = 0$ c) $2x^2 - 10x + 3 = 0$ d) $x^2 - 12x - 7 = 0$

Graphs in practical situations

Conversion graphs

Conversion graphs allow you to convert between two different measures, such as currency exchange, miles to kilometres or temperature in °F and °C.

If you are asked to draw a graph from given data you will need to decide if the line passes through the origin (0, 0): £0 = $0, for example, but 0°C = 32°F.

This graph has been drawn to show the connection between the mass of a turkey (horizontal axis) and its cooking time (vertical axis).

To find the time to cook a 6 kg turkey draw a line up from 6 kg to meet the conversion line and then across to read the hours needed, 3.5 hours (3 hours 30 minutes).

To find the mass of a turkey with a cooking time of 5 hours, draw a line across from 5 hours until it meets the conversion line and then down to find the mass, 10.5 kg.

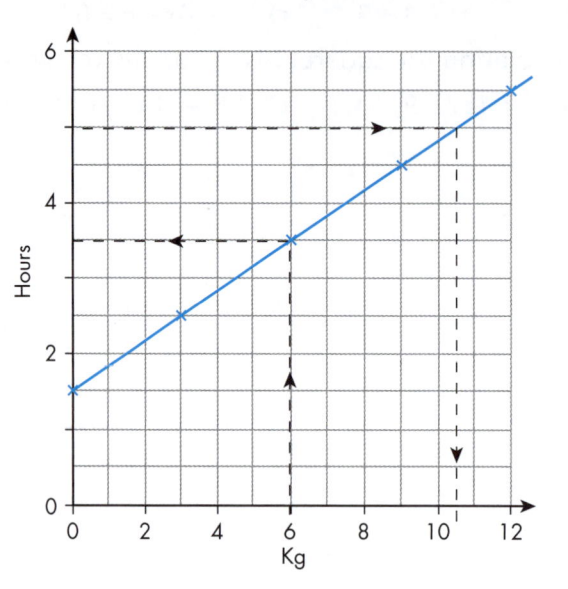

Travel graphs (distance–time graphs)

Travel graphs, or **distance–time graphs**, give information about a journey. You can read information from them by drawing lines from one axis to meet the graph line and then to meet the other axis.

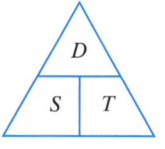

This graph shows the journeys of Hannah and Alice who are travelling 40 km to a destination. Hannah sets off at 8 a.m. on her bicycle. At 9 a.m. she has a rest for 15 minutes and then sets off again. Alice sets off at 9.30 a.m. in her car and drives at a constant speed. Alice passes Hannah at a few minutes after 10 a.m.

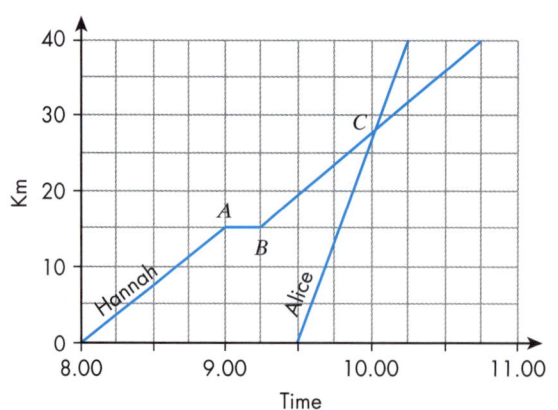

Hannah travelled 15 km in the first hour, so her speed was 15 km/h.

For the second part of her ride Hannah took $1\frac{1}{2}$ hours to cover 25 km, $\frac{25}{1.5}$ = 16.7 km/h (3 s.f.).

> **Revision tip**
>
> In travel graphs, a horizontal line indicates no movement (rest) and a diagonal straight line represents constant speed.

Extended

Speed–time graphs

A speed–time graph shows different information to a travel graph. On a travel graph a horizontal line indicates no movement (rest). On a speed–time graph a horizontal line represents constant speed. A diagonal line represents change in speed. The gradient of the line represents acceleration. Positive gradient shows increase in speed, negative gradient shows decrease in speed, this is called deceleration.

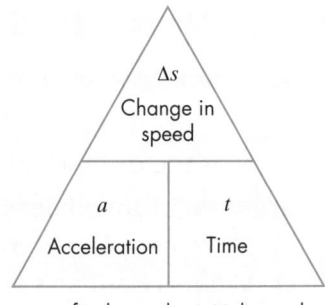

$$\text{acceleration} = \frac{\text{change in speed}}{\text{time taken}}$$

The solid line on this graph shows the speed of a car as it moves between two sets of traffic lights. The dashed line represents the speed of a cyclist.

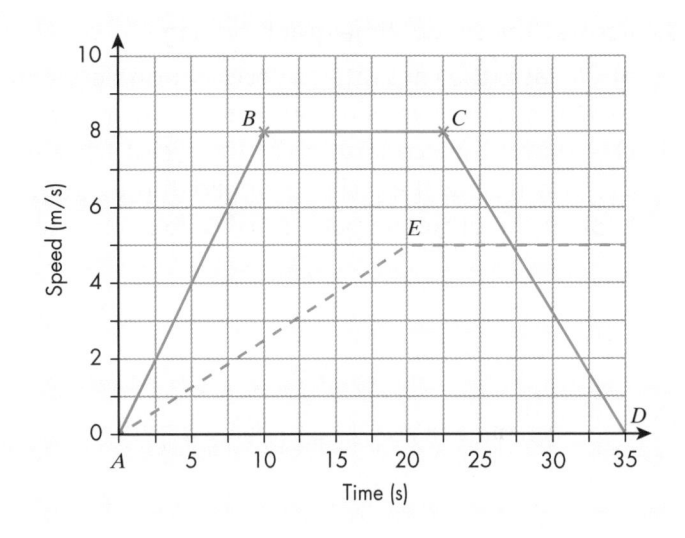

The car accelerates for 10 seconds, travels for 12.5 seconds at 8 m/s and then slows down to a stop. The total journey is 35 seconds.

The car's acceleration over the first 10 seconds:

$$\frac{8-0}{10} = \frac{8}{10} = 0.8 \text{ m/s}^2$$

The area under the graph represents the total distance travelled. This graph is a trapezium.

$$\text{area} = \frac{1}{2}(12.5 + 35) \times 8 = 190 \text{ m travelled}$$

If the rate of acceleration or deceleration is not constant the lines will be curves.

Milo is running in a race at school sports day. The graph of his first 15 seconds might look like this:

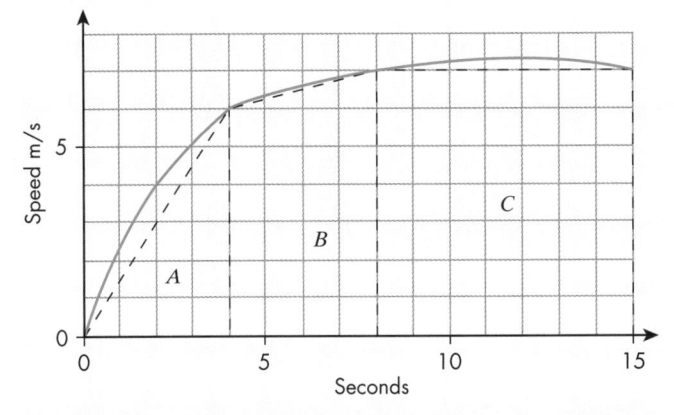

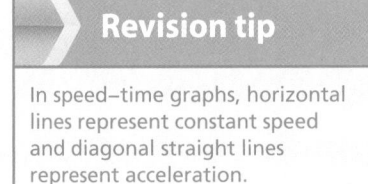

> **Revision tip**
>
> In speed–time graphs, horizontal lines represent constant speed and diagonal straight lines represent acceleration.

He sprints out of the blocks and gradually gets faster but soon reaches a maximum speed and then slows down a little before getting into a comfortable stride. To find how far he has run in 15 seconds break the graph into areas that you can work out – triangles, rectangles and trapezium – and add them together:

$$\text{area} = A + B + C = \left[\frac{1}{2}(4 \times 6)\right] + \left[\frac{1}{2}(6+7) \times 4\right] + [7 \times 7] = 87 \text{ m}$$

Note that this is an *under-estimate* of the actual distance because there are some small areas between the shapes created and the curve itself that are not included.

The gradient at a point on a speed-time graph gives the acceleration at that point. You can find the gradient at a particular point by drawing a tangent to the graph at that point, this is explained fully under 'Estimating gradients' on page 65.

Quick test

1. Look at the conversion graph on page 56.
   a) A turkey needs 4 hours 30 minutes to cook. What is its mass?
   b) Can the graph be used to estimate the time needed to cook a 20 kg turkey? Justify your answer.
2. Look at the distance–time graph on page 56.
   a) At what speed does Alice travel the journey (to the nearest km/h)?
   b) How many km from the start of their journeys did Hannah pass Alice?
   c) What is Hannah's average speed for the total journey (to the nearest km/h)?

Extended

3. Look at the speed–time graph on page 57 of a car moving between two sets of traffic lights.
   a) What is the car's rate of deceleration between points C and D?
   b) What is the average speed of the car over the 35 seconds, in km/h (to 1d.p.)?
   c) What is the acceleration of the cyclist from A to E?
   d) How far has the cyclist travelled in the 35 seconds?
4. Look at the graph on page 57 of Milo's race. How far has he run in the first 10 seconds?

Straight-line graphs

Using coordinates

A graph space, or grid, has four quadrants, with the top right section being the first quadrant and the others counted anticlockwise. Coordinates are given in the form (x, y).

Vertical and horizontal graph lines are identified by only an x- or a y-value. $x = 2$ (vertical) and $y = -1$ (horizontal) are marked on the diagram in blue. The x-axis has the equation $y = 0$ and the y-axis has the equation $x = 0$

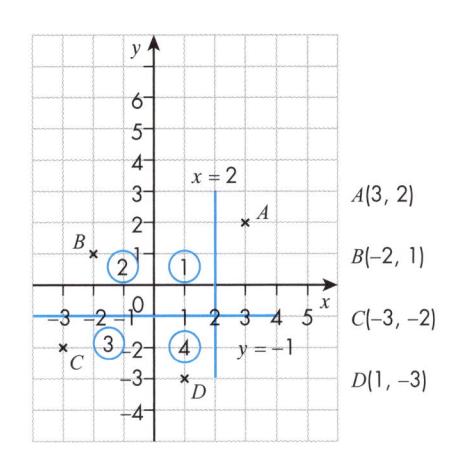

$A(3, 2)$
$B(-2, 1)$
$C(-3, -2)$
$D(1, -3)$

Straight-line graphs

If an equation has the form $y = mx + c$, with m and c being constants, it will produce a straight line when plotted.

x	-2	0	5
y	-8	-2	13

The diagram shows the graph $y = 3x - 2$ for values of x in the range $-2 \leq x \leq 5$. The small grid is called a table of values. You may be given a grid like this table to complete, with the x-values provided.

If you are not given the table, choose a value for x, such as $x = 5$, and work out the equivalent y-value.

$y = 3(5) - 2 = 15 - 2 = 13$

This gives the coordinates $(5, 13)$.

> **Revision tip**
>
> Always complete a table of values with the minimum and maximum values of x first so that you will then know how long to make/draw the y-axis.

You need a minimum of two coordinates in order to plot a straight-line graph. However, plotting a third point provides a check to ensure that you do create a straight line. Calculate values of y for three x-values and plot them on the given graph space.

Always label the line that you have plotted by neatly writing the equation of the line at a convenient place on the line.

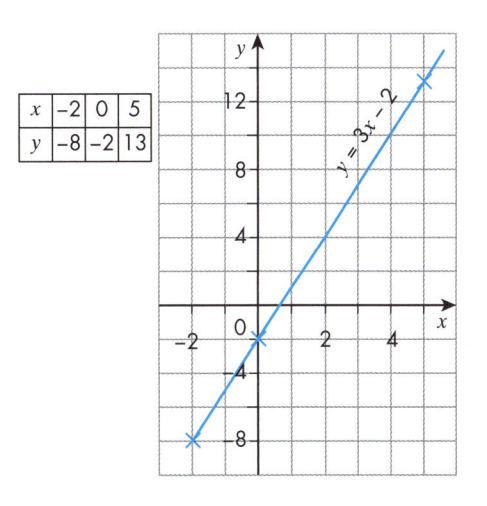

The equation $y = mx + c$

Gradient is a measure of how steep a line is.

The gradient of a line joining two points is calculated as: $\frac{\text{change in } y\text{-values}}{\text{change in } x\text{-values}}$, counted along the axes, and results in a number without units.

Look at the graph of $y = 3x - 2$, where a small triangle has been drawn on the line. Imagine you are going to walk up the slope. The change in the value of y would be 6 and of x it would be 2. The gradient is $\frac{6}{2} = 3$

The gradient is represented by m in $y = mx + c$.

The value of the constant c identifies the point where the line crosses the y-axis (the y-**intercept**), in this case -2

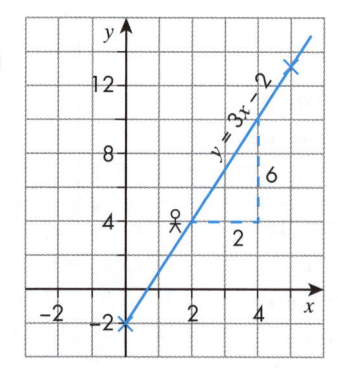

You can see that these values can be identified in the equation $y = 3x - 2$

The equation of a line

Using these ideas, it is possible to work out the equation of a given line.

In this diagram the change in y from the point you are walking down the slope is -6 and the change in x is 3. Note that if you started walking at the bottom of the slope the change in y is $+6$ and in x is -3.

The gradient is $\frac{-6}{3} = -2$. The intercept is 10

The equation of the line is $y = 10 - 2x$

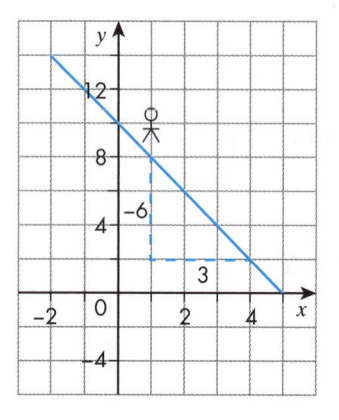

Parallel lines

Look at the equations of the four lines in this diagram.

The lines are all **parallel** and the equations all share a common element — the gradients are the same.

Any lines that are parallel have the same gradient.

Imagine drawing the line for the equation $y = 2x + 5$. Then think about drawing another line parallel to this, but passing through the point with coordinates (5, 6). What is the equation of this second line?

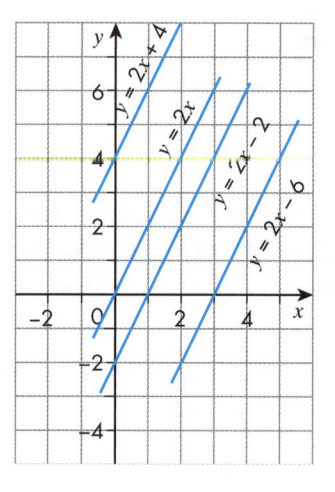

The line will have a gradient of 2, so has the equation $y = 2x + c$.

Substitute (5, 6) into the equation: $6 = 2(5) + c \rightarrow c = -4$

So the equation is $y = 2x - 4$.

Extended

Midpoints of a line

The midpoint of a line segment is the point on that line segment that is the same distance from each end point.

To find the coordinates of the midpoint, add the x-coordinates of the end points and divide by 2, and do the same for the y-coordinates.

The coordinates of the midpoint of the line from (x_1, y_1) to (x_2, y_2) are $\left(\dfrac{x_1+x_2}{2}, \dfrac{y_1+y_2}{2}\right)$.

The midpoint of the line on the right has coordinates $\left(\dfrac{-2+3}{2}, \dfrac{3+(-1)}{2}\right)$ i.e. (0.5, 1)

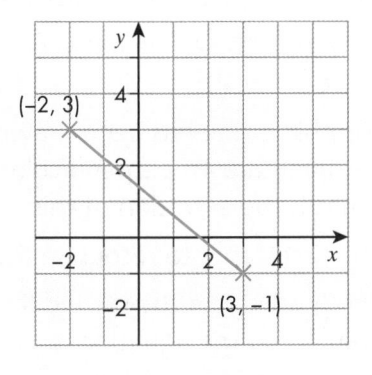

The distance between two points

You can use Pythagoras' theorem ($a^2 + b^2 = c^2$) to find the length of a line segment.

In the same way as you would work out the gradient, draw a triangle to find the changes in the x-values and the y-values, which are 5 and 4

The length of the line is $\sqrt{5^2 + 4^2} = \sqrt{41} = 6.40$ (2 d.p.).

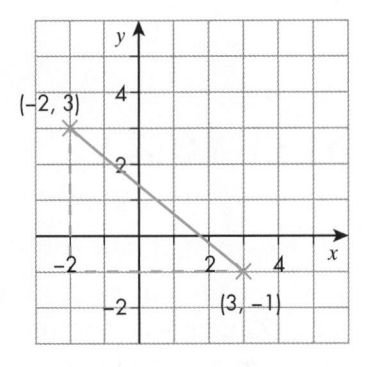

The equation of a line segment

If a line segment does not cross the y-axis it is not possible to read the value of c. But, it is still possible to work out the gradient of the line, so a value for m can be found. You will know the coordinates of the end points of the line segment, so these can be used, together with the gradient, to find a value for c and hence the equation of the line segment.

For example, in the diagram the line segment AB has end coordinates of (2, 3) and (6, 5).

The gradient of AB is $\dfrac{2}{4} = 0.5$

This gives $y = 0.5x + c$

You know that the line passes through (2, 3) so substitute these into $y = 0.5x + c$, and solve for c:

$3 = 0.5(2) + c$

$3 - 1 = c$

$2 = c$

This gives an equation for AB: $y = 0.5x + 2$

(Note that using the coordinates (6, 5) would also give a value $c = 2$)

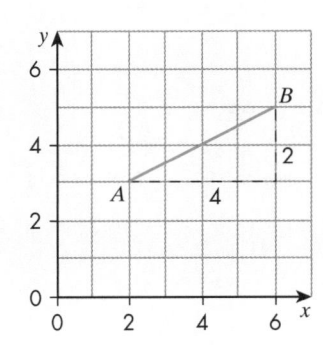

Perpendicular lines

In this diagram lines A and B are **perpendicular** to each other, that is, the angle between them is 90°.

You can see from the triangles drawn with dashed lines that the gradient of A is $\dfrac{4}{-2} = -2$, and the gradient of B is $\dfrac{1}{2} = 0.5$

The relationship between these is that the gradient of A = –(the reciprocal of the gradient of B)

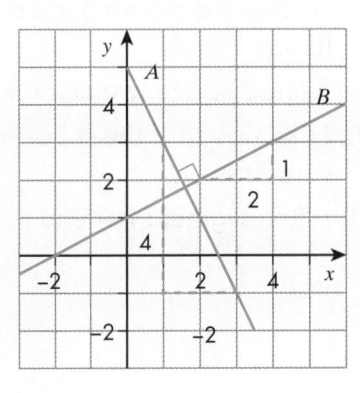

Algebra

The gradient of $B = 0.5$, so the gradient of $A = -\frac{1}{0.5} = -2$.

Alternatively, if the gradient of $A = -2$, the gradient of $B = -\frac{1}{-2} = 0.5$

Note that the angle between perpendicular lines will only appear to be 90° if the scales on the x- and y-axes are the same.

> **Revision tip**
>
> Remember that lines with a positive gradient rise from left to right and lines with negative gradients fall from left to right.

Suppose a line has been drawn with the equation $y = 3x - 5$. Another line has been drawn perpendicular to this and passes through the point (3, 6). What is the equation of this second line?

The first step is to recognise the gradient of the original line, in this case, 3.

The gradient of the second line is the negative reciprocal of the original line, $-\frac{1}{3}$, so the equation is $y = -\frac{1}{3}x + c$.

Substitute (3, 6) into this: $6 = -\frac{1}{3}(3) + c$, so $c = 7$, which gives an equation of $y = -\frac{1}{3}x + 7$.

> **Revision tip**
>
> The products of the gradients of perpendicular lines will always equal –1, unless one of the lines has a gradient of 0.

Quick test

Use this diagram to answer these questions.

1. What are the gradients of lines AB and CD?
2. The equation of PQ is $y = x + 6$. Another line RS is parallel to PQ and passes through point (3, 2). What is its equation?
3. The equation of VW is $y = 14 - 3x$. A line TU is parallel to VW and passes through point (–2, 6). What is its equation?
4. What are the equations of lines AB and CD?

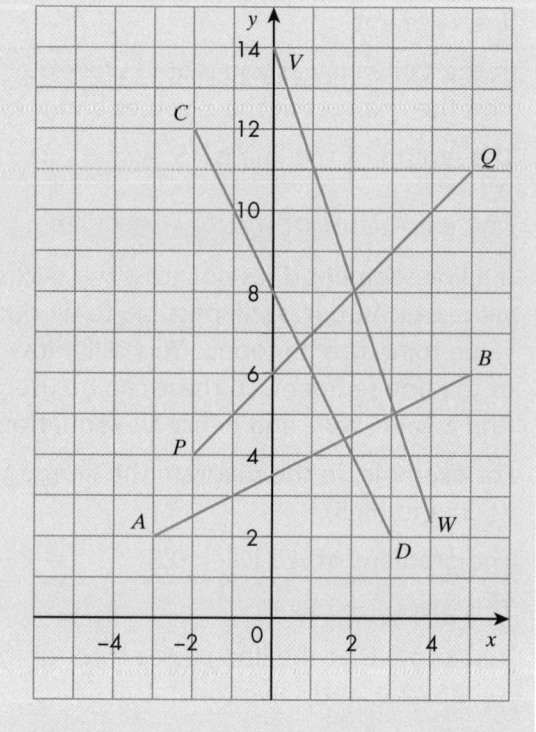

Extended

5. What are the coordinates of the midpoints of AB and CD?
6. A line EF is drawn parallel to line AB. It passes through point (10, 2). What is its equation?
7. A line GH is drawn parallel to line CD. It passes through point (–4, 5). What is its equation?
8. A line JK is drawn perpendicular to line AB. It passes through the point (–4, 4). What is its equation?
9. A line LM is drawn perpendicular to line CD. It passes through the point (6, 2). What is its equation?
10. Erika says that the shape made by lines AB, CD, JK and LM is a rectangle. Is she right? Justify your answer.
11. A line segment has end coordinates of (2, 7) and (5, 1). What is the equation of this line segment?

Graphs of functions

Quadratic graphs

The graph of a quadratic function, one that is of the form $y = ax^2 + bx + c$, will always be a smooth **parabola**.

When drawing a quadratic graph you should always draw a table of values to find a set of coordinates to plot.

To draw the graph of $y = x^2 - 4x + 1$ for $-1 \leq x \leq 5$, first draw a table of values, splitting the equation into its individual terms.

x	−1	0	1	2	3	4	5
x^2	1	0	1	4	9	16	25
$-4x$	4	0	−4	−8	−12	−16	−20
$+1$	1	1	1	1	1	1	1
$y = x^2 - 4x + 1$	6	1	−2	−3	−2	1	6

> **Revision tip**
>
> The roots (solutions) of an equation are where the curve crosses the x-axis, i.e. where $y = 0$.

The top and bottom rows provide the coordinates to plot. Remember to draw a smooth curve through all of the points. Do not join the points with a ruler.

You can use the graph to answer questions.

a) What are the values of x when $y = 3$?

  Draw the line $y = 3$ and read off the values for x: −0.4 and 4.4

b) What is the value of y when $x = 1.5$?

  Draw a line $x = 1.5$ and read off the corresponding value of y, −2.8

c) What are solutions for $x^2 - 4x + 1 = 0$?

  This requires the **roots** of the equation, when $y = 0$, $x = 0.3$ or $x = 3.7$

d) Where is the **turning point** of the graph?

  This is the lowest, or **minimum point**: (2, −3)

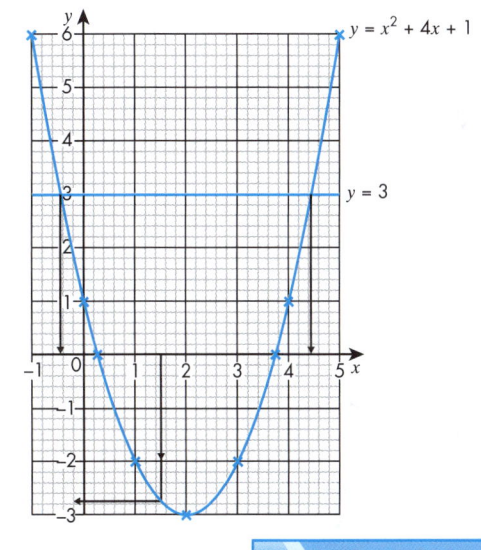

> **Revision tip**
>
> You need to be able to pick out the significant points of the curve: roots, y-intercept and turning point.

Extended

On page 54 you were shown how to solve a quadratic equation by 'completing the square'.

Look at the quadratic that started this section: $x^2 - 4x + 1$

If this is rearranged by completing the square it gives: $(x - 2)^2 - 3$

From the graph of this quadratic you can see that the turning point was identified as (2, −3). These coordinates can be found from $(x - 2)^2 - 3$ with the x coordinate being the value in the bracket with its sign reversed and the y coordinate being the value outside the bracket.

This method can be used to determine the turning point of any quadratic curve.

By solving $(x - 2)^2 - 3 = 0$ you can also find the roots: $x = 2 \pm \sqrt{3}$, 3.7 and 0.27

Reciprocal graphs

The reciprocal equation is of the form $y = \frac{a}{x}$

You can plot and draw a graph for $y = \frac{1}{x}$ from these values.

Number (x)	1	2	3	4	5	7	8	10
Reciprocal (y)	1	0.5	0.33	0.25	0.2	0.14	0.125	0.1

Plotting the graph for values of $x < 0$ produces a matching graph in the third quadrant.

The graph of the reciprocal equation has a very distinct shape and properties.

The closer x gets to zero the closer the graph gets to the y-axis. As x increases, the closer the graph gets to the x-axis.

The graph never touches either of the axes. The x- and y-axes in this case are known as **asymptotes**, as the graph never crosses the axes but gets closer and closer to them.

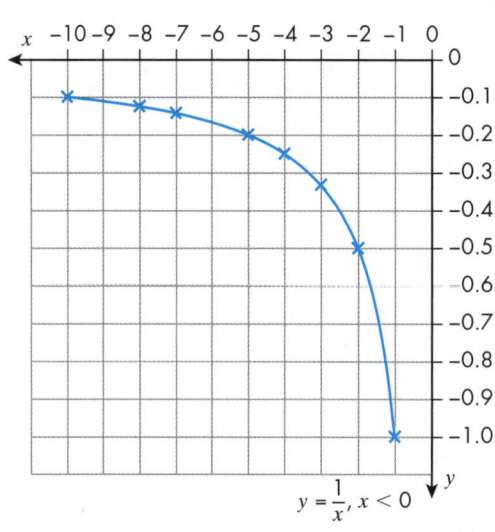

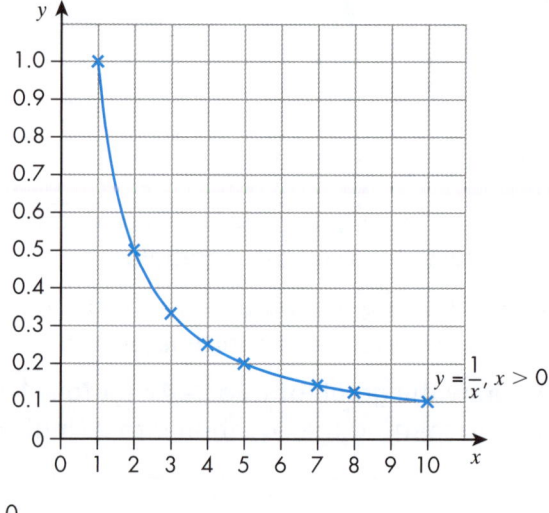

Extended

Cubic graphs

The graph of a cubic function, one that is of the form $y = ax^3 + bx^2 + cx + d$, is a smooth curve that has two changes of direction.

To draw the graph of: $x^3 - x^2 - x + 1$, for values of x: {−1.5, −1, −0.5, 0, 1, 2}, start with a table of values to find the coordinates of the points to be plotted.

x	−1.5	−1	−0.5	0	1	2
$y = x^3 - x^2 - x + 1$	−3.125	0	1.125	1	0	3

Once you have drawn the graph, you can use it to answer questions.

What are the values of x when $y = 0.5$?

Draw in the line $y = 0.5$ and then read off the values for x: −0.85, 0.4, 1.3

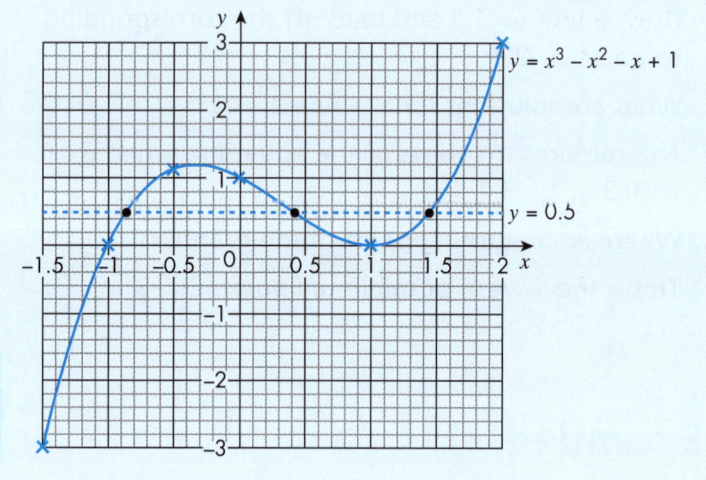

Exponential graphs

Exponential equations have a term in which the variable is an index.

> **Revision tip**
>
> In exponential graphs, for values of $x > 0$, as x increases, the value of y increases steeply. For values of $x < 0$ as x decreases the line of the graph gets closer to the x axis but never falls below it.

Example

Zac started a new Twitter account 6 months ago and currently has 150 followers. He started off with 16 followers who were work colleagues and friends. After recording his number of followers for the first 6 months, he worked out that very roughly his followers increase in number at a rate of 45% a month.

Zac came up with the calculation 150×1.45^x (where x is the month number from now) to model how his follower numbers will increase.

x	1	2	3	4	5	6
$y = 150 \times 1.45^x$	218	315	457	663	961	1394

The resulting graph is a curve that crosses the y-axis at the starting value (150) and as it continues left into the second quadrant, back through the previous months' follower figures, the curve will approach Zac's starting value of 16 followers.

When an exponential graph is extended for values of $x < 0$, it gets closer to the x-axis but, as with the reciprocal graph, never meet it.

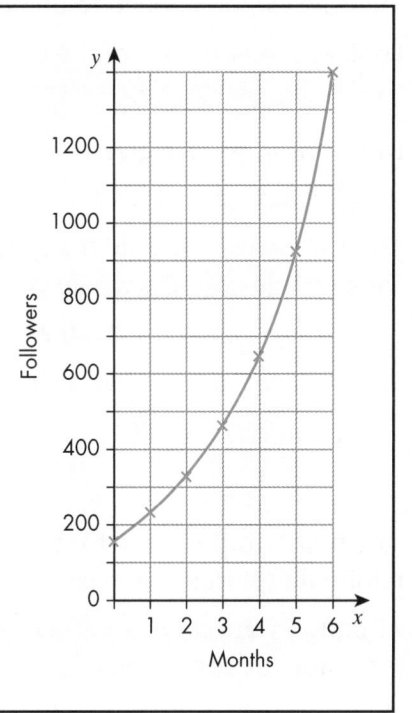

Estimating gradients

The gradient of a curve changes all along the length of the curve. It is possible to estimate the gradient of the curve at a particular point by drawing a **tangent** to the curve at that point.

Then you can find the gradient of the tangent in the same way as you find the gradient of a straight-line.

In the diagram, tangents have been drawn at points A and C.

Triangles have been drawn, using the tangents, so the gradients can be calculated.

The gradient at point A is $\dfrac{-5.5}{2.5} = -2.2$

The gradient at point C is $\dfrac{3}{4} = 0.75$

The gradients at points B and D are zero.

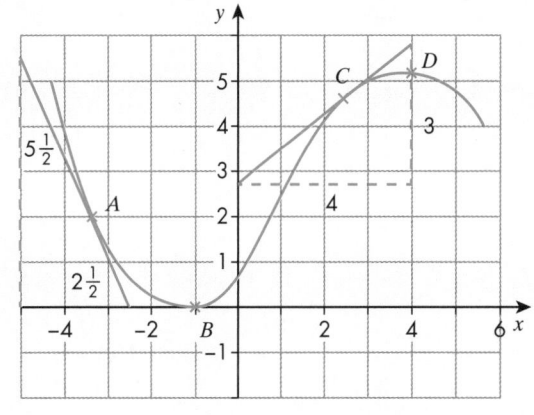

Differentiation

Differentiation is an algebraic method of accurately finding the gradient of a curve, without having to draw a tangent to the curve first.

You need to find the derivative of y with respect to x.

This is written as $\dfrac{dy}{dx}$

Differentiation (continued)

The rule for differentiating is:

If $y = x^n$, then $\frac{dy}{dx} = nx^{n-1}$

In its simplest case this would mean that if $y = x^3$, $\frac{dy}{dx} = 3x^2$

The resulting $\frac{dy}{dx}$ expression is known as a *derived function* because it is a derivative of the original function.

The rule can be applied to any term in an expression.

Look at the equation of the curve in the example from page 64:

$y = x^3 - x^2 - x + 1$

Differentiating this gives:

$\frac{dy}{dx} = 3x^2 - 2x - 1$ $[x^2 \rightarrow 2x, x \rightarrow 1$ (remember $x = x^1$), $1 \rightarrow 0]$

This can be used to find the gradient at any point x. So the gradient at, for example, $x = -0.5$ would be:

$3(-0.5)^2 - 2(-0.5) - 1 = 0.75$

The gradient of the curve at the turning point where $x = 1$ should be zero:

$3(1)^2 - 2(1) - 1 = 0$

Maxima and minima

Any point on the curve that is a turning point is either a maximum or minimum (plurals are maxima and minima).

Visually, in the curve $y = x^3 - x^2 - x + 1$, you can see that there is a maximum at approximately -0.35 and a minimum at $+1$

You can confirm this using algebra: $\frac{dy}{dx} = 3x^2 - 2x - 1$

For turning points $\frac{dy}{dx} = 0$

So $3x^2 - 2x - 1 = 0$

you get $x = -\frac{1}{3}$ and $+1$ – these are the x-coordinates of the turning points.

Quick test

1. Draw the graph of $y = x^2 - 4x - 1$ for $-1 \leqslant x \leqslant +5$.
   a) What are the coordinates of the turning point?
   b) What are the values of x when $y = 3$?
   c) What is the value of y when $x = 2.5$?
   d) What are the solutions for $x^2 - 4x - 1 = 0$

2. a) Copy and complete the table of values for $y = \frac{6}{x}$ for $1 \leqslant x \leqslant 12$

x	1	2	3	4	6	8	12
y							

   b) Draw the graph of $y = \frac{6}{x}$
   c) On the same axes, draw the line $y = x - 1$
   d) Use your graph to solve the equation $\frac{6}{x} = x - 1$

Extended

3. The capacity, in millilitres, of a glass bowl is $\frac{1}{12}\pi x^3$, where x is the diameter in centimetres.
   a) Copy and complete the table of values for $y = \frac{1}{12}\pi x^3$. Give your answers to 2 s.f.

x	15	16	17	18	19	20
y		1100				2100

   b) Draw the graph of $y = \frac{1}{12}\pi x^3$
   c) Use your graph to find the diameter of a bowl with a capacity of 1400 ml.

4. The value of a piece of farm machinery is depreciating by 30% per year. The value now is $20 000.
   a) Copy and complete the table to find its value in future years (to 2 s.f.).

x	0	1	2	3	4	5
y	20 000				4800	

   b) Draw a graph to show how the value of the machinery varies over these 6 years.
   c) After how many complete years will the machinery have dropped below half its current value?
   d) At approximately what age will the machinery be worth $4000?

5. a) Look at the graph for $y = x^3 - x^2 - x + 1$ on page 64. What is the gradient of the curve where $x = -0.5$?
   b) Look at the graph for $y = \frac{1}{x}$ when $x > 0$, on page 64. What is the gradient of the curve where $x = 2$?

6. By completing the square, find the turning point and roots (leaving in surd form) of $x^2 + 4x - 7$

Algebra

Number sequences

Patterns in number sequences

A **number sequence** is an ordered set of numbers that are connected by a rule. Knowing the rule allows you to generate numbers in the **sequence**.

Each number in the sequence is called a term.

4, 10, 16, 22, 28, …	The rule is 'add 6'. Start at 4 and then add 6 each time, so the next two terms are 34 and 40.
28, 25, 22, 19, …	Start at 28 and then subtract 3 each time, so the next two terms are 16 and 13.
512, 256, 128, 64, …	The rule is 'divide by 2'. Start at 512 and then divide by 2 each time, so the next two terms are 32 and 16.

The pattern in which each term, apart from the first one, is derived from the term before it is called the term-to-term rule.

Differences

Sometimes the differences between the terms will help you to find the connection between term.

Find the next two items of the sequence.

5   7   11   17   25 . . . .

Look at the differences.

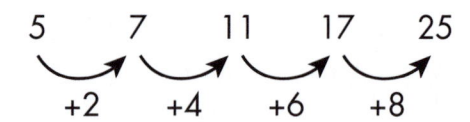

5   7   11   17   25
 +2   +4   +6   +8

Each time the difference increases by 2

So the sequence continues:

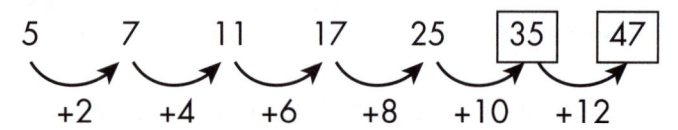

5   7   11   17   25   35   47
 +2   +4   +6   +8   +10   +12

> **Revision tip**
>
> Sometimes the differences between the terms form a sequence of their own so you may need to work out this sequence before expanding the original sequence.

The nth term of a sequence

Knowing the rule that connects the terms in the sequence is helpful until you are asked to find, for example, the 50th or 100th term in the sequence. You are not expected to write a long list of 100 terms to find the answer.

You should use an expression, called the **nth term**, which is the general rule for the sequence.

If the nth term of a sequence is $4n - 1$, where n is the term number (or position), the first five terms of the sequence will be:

n	1	2	3	4	5
Term	$4(1) - 1$	$4(2) - 1$	$4(3) - 1$	$4(4) - 1$	$4(5) - 1$
	3	7	11	15	19

So the sequence is: 3, 7, 11, 15, 19, …

The 50th and 100th terms would be $4(50) - 1 = 199$ and $4(100) - 1 = 399$

Finding the nth term of a linear sequence: $An + b$

In a **linear sequence** the difference between one term and the next is always the same.

The general form for the nth term of a linear sequence is $An + b$.

In the sequence: 4, 10, 16, 22, 28, … the difference is 6

A, the coefficient of n, is the difference between the terms, so $A = 6$

b is the difference between the first term and A, so $b = 4 - 6 = -2$

The nth term is $6n - 2$

The 50th term in this sequence will be: $6(50) - 2 = 298$

Triangular numbers

This is a special sequence of numbers which follows the pattern:

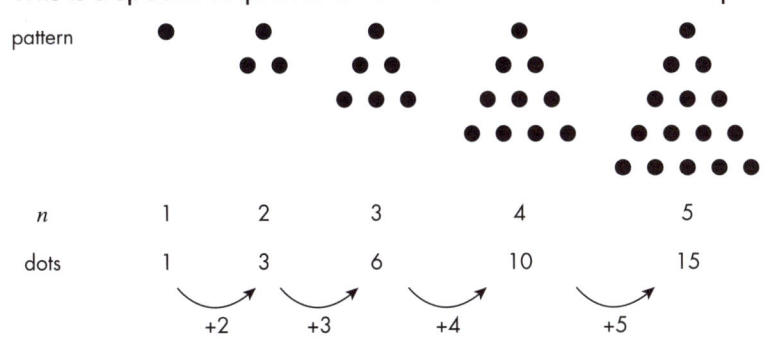

n	1	2	3	4	5
dots	1	3	6	10	15

(+2, +3, +4, +5)

The nth term for finding triangular numbers is, $n(n + 1) \div 2$

Quadratic and cubic sequences

The square numbers are: 1, 4, 9, 16, 25, … so the nth **square number** is n^2.

A sequence based on square numbers is a quadratic sequence.

The cube numbers are: $1^3, 2^3, 3^3, 4^3, 5^3$, … which gives 1, 8, 27, 64, 125, … so the nth cube number is n^3.

A sequence based on cube numbers is a cubic sequence.

In the sequence: −2, 1, 6, 13, 22, … the differences are not the same, so the sequence is not linear.

Compare the sequence with the square numbers…

$\quad$ −2 $\quad$ 1 $\quad$ 6 $\quad$ 13 $\quad$ 22

$\quad$ 1 $\quad\;$ 4 $\quad$ 9 $\quad$ 16 $\quad$ 25

… and you can see that each term of the sequence is 3 less than the corresponding square number.

The nth term is: $n^2 − 3$.

In the sequence: 10, 17, 36, 73, 134, … the differences are not the same, so the sequence is not linear.

Compare the sequence with the square and cube numbers…

$\quad$ 10 $\quad$ 17 $\quad$ 36 $\quad$ 73 $\quad$ 134

$\quad$ 1 $\quad\;$ 4 $\quad\;$ 9 $\quad$ 16 $\quad$ 25

$\quad$ 1 $\quad\;$ 8 $\quad$ 27 $\quad$ 64 $\quad$ 125

… and you can see that each term of the sequence is 9 more than the corresponding cube number.

The nth term is: $n^3 + 9$

General rules from patterns

You may be given a sequence of diagrams from which to work out a pattern, discovering the nth term and making a prediction.

In the diagram, hexagonal tables at a primary school can each seat six pupils, but may be put together as shown.

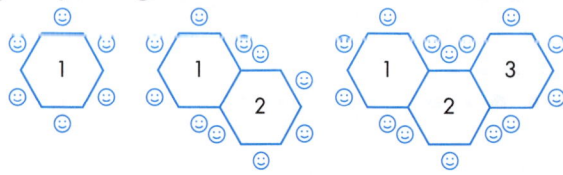

> **Revision tip**
>
> When finding a general rule from a sequence of diagrams (seats, bricks, squares…), always set up a table that connects the pattern number with the term of the sequence.

How many pupils could sit at five tables put together in this way?

How many pupils could be seated at 20 connected tables?

Draw a couple more patterns and then put the information in a table.

Tables	1	2	3	4	5	6
Number of pupils	6	10	14			

To find the nth term:

- the difference in consecutive numbers of pupils from one arrangement to the next is 4
- the difference between the first number and this difference is 6 − 4 = 2
- so the nth term is $4n + 2$.

The number of pupils sitting at 20 tables is $4 \times 20 + 2 = 82$.

Extended

Exponential sequences

The nth term of an exponential sequences has the value of n as an index number.

The nth term 1.25×4^n generates the sequence: 5, 20, 80, 320, 1280, ...

You multiply each term by 4 to find the next one. Dividing the first term (5) by the multiplier (4) gives 1.25

Subscript notation

It is useful to be able to refer specifically to a term in the sequence and in the sections on the previous pages you have been asked, for example, to 'Find the 50th term of the sequence'. Instead, this could be rewritten as, 'Find x_{50}'. The subscript number refers to the particular term in the sequence.

For example: the nth term of a sequence is found using the formula $x_n = 4n - 3$

Find the value of   **a)** x_5 **b)** x_{50}

a) $x_5 = 4(5) - 3 = 17$ **b)** $x_{50} = 4(50) - 3 = 197$

Quick test

1. Given these nth terms, write down the first five terms for each sequence.
   - **a)** $n + 3$
   - **b)** $2n - 3$
   - **c)** $3n^2 + 2$
   - **d)** $24 - n^2$

2. Find the next three terms and the nth term for each linear sequence.
   - **a)** 11, 18, 25, 32, 39, ...
   - **b)** 3, 7, 11, 15, 19, ...
   - **c)** 67, 58, 49, 40, 31, ...
   - **d)** 19, 13, 7, 1, –5, ...

3. This is the start of a series of patterns made from matchsticks.
   - **a)** Copy the sequence and draw the fourth and fifth patterns.
   - **b)** What is the nth term of this sequence?
   - **c)** How many matches are needed for the 35th term in the sequence?
   - **d)** What is the largest pattern that you could make with 300 matches?

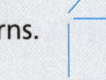

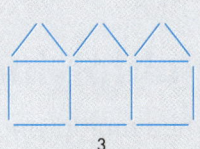

Extended

4. Find the next term in each of these exponential sequences.
   - **a)** 3, 12, 48, 192, ...
   - **b)** 7.5, 22.5, 67.5, 202.5, ...

5. Work out the nth term of each exponential sequence.
   - **a)** 9, 54, 324, 1944, ...
   - **b)** 6, 18, 54, 162, ...

6. If $x_n = n^2 + 2n - 5$, find:
   - **a)** x_5
   - **b)** x_{20}

Algebra

Indices

You have already used indices (or index numbers) in the sections on squares and cubes, and standard form.

In 5^2 and 3×10^7, the numbers 2 and 7 are indices, or powers.

The index number indicates how many 'lots' of a number (or variable) are multiplied together.

$m^4 = m \times m \times m \times m$, and $4^6 = 4 \times 4 \times 4 \times 4 \times 4 \times 4$

To enter an index number on your calculator use the $x^\blacksquare$ button on your calculator.

$4^6 =$ [4] [$x^\blacksquare$] [6] = 4096

There are two special cases to remember.
- A number to the power of 1 is the number itself, $4^1 = 4$, $x^1 = x$
- A number to the power of 0 is equal to 1, $4^0 = 1$, $x^0 = 1$

Negative indices

Writing a term with a negative index is a more convenient way of writing the reciprocal of that term.

So, $m^{-3} = \frac{1}{m^3}$ and $6^{-3} = \frac{1}{6^3}$

Also, $\frac{1}{t^3} = t^{-3}$ and $\frac{1}{4^5} = 4^{-5}$

> **Revision tip**
>
> A negative power means a reciprocal – it does not mean that the answer is negative.

Multiplying and dividing with indices

When multiplying powers of the same number or variable, add the indices.

$3^2 \times 3^4 = 3^6$ $m^7 \times m^3 = m^{10}$ $n^2 \times n^3 \times n^6 = n^{11}$ $p^6 \times p^{-4} = p^2$

When dividing powers of the same number or variable, subtract the indices.

$5^4 \div 5^2 = 5^2$ $q^5 \div q^{-2} = q^7$ $s^{-3} \div s^{-5} = s^2$ $t^3 \div t^8 = t^{-5}$

When raising a power to a further power, multiply the indices.

$(4^3)^4 = 4^{12}$ $(v^4)^{-2} = v^{-8}$ $(w^{-2})^{-4} = w^8$

When expressions include a mixture of numbers and variables remember to multiply or divide the numbers as you would in any other expression.

$4c^2 \times 5c^3 = (4 \times 5) \times (c^2 \times c^3) = 20c^5$

$4d^3e^2 \times 3d^2e^4 = (4 \times 3) \times (d^3 \times d^2) \times (e^2 \times e^4) = 12d^5e^6$

$18g^5 \div 6g^3 = (18 \div 6) \times (g^5 \div g^3) = 3g^2$

$(3h^2)^3 = (3)^3 \times (h^2)^3 = 27h^6$

> **Revision tip**
>
> Write your solution with the number first and then the letters in alphabetical order.

IGCSE Mathematics Revision Guide

Fractional indices

Fractional indices may take one of two forms: $x^{\frac{1}{n}}$ or $x^{\frac{a}{b}}$, and either of these may be positive or negative.

Indices of the form $x^{\frac{1}{n}}$ follow the pattern $x^{\frac{1}{n}} = \sqrt[n]{x}$, where n describes the root of x.

This means that, for example, $x^{\frac{1}{2}} = \sqrt{x}$ (the square root) and $x^{\frac{1}{3}} = \sqrt[3]{x}$ (the cube root).

If the index is a negative fraction then you need the reciprocal of the fractional index: $x^{-\frac{1}{4}} = \frac{1}{\sqrt[4]{x}}$

If $\left(\frac{2}{3}\right)^2 = \frac{2^2}{3^2} = \frac{4}{9}$ then, in the same way, $\left(\frac{36}{49}\right)^{\frac{1}{2}} = \frac{36^{\frac{1}{2}}}{49^{\frac{1}{2}}} = \frac{\sqrt{36}}{\sqrt{49}} = \frac{6}{7}$

Indices of the form $x^{\frac{a}{b}}$ follow the pattern $x^{\frac{a}{b}} = \sqrt[b]{x^a}$

$w^{\frac{3}{4}} = \sqrt[4]{w^3}$

$t^{\frac{5}{2}} = \sqrt{t^5}$

$\left(\frac{c}{d}\right)^{-\frac{3}{4}} = \left(\frac{d}{c}\right)^{\frac{3}{4}}$

$27^{\frac{2}{3}} = \left(\sqrt[3]{27}\right)^2 = 3^2 = 9$

Quick test

1. Evaluate these numbers.
   a) 5^4
   b) 4^{-2}
   c) 6^1
   d) 8^0

2. Simplify these expressions.
   a) $3^5 \times 3^2$
   b) $5^{-3} \div 5^{-5}$

3. Evaluate these numbers.
   a) $81^{\frac{1}{2}}$
   b) $25^{\frac{3}{2}}$
   c) $256^{-\frac{3}{4}}$
   d) $\left(\frac{8}{125}\right)^{\frac{2}{3}}$

Extended

4. Write these in fraction form.
   a) $3a^{-2}$
   b) $6a^{-2}$
   c) $\frac{1}{2}a^{-2}$
   d) $\frac{4}{5}b^{-3}$

5. Write these in index form.
   a) $\frac{5}{v^2}$
   b) $\frac{3}{w^6}$
   c) $\frac{2}{t^4}$
   d) $\frac{z}{a^2}$

6. Simplify these expressions.
   a) $2m^8 \times 3m^{-5}$
   b) $12n^4 \div 3n^{-2}$

7. Simplify these expressions:
   a) $e^{\frac{1}{2}} \times e^{\frac{1}{3}}$
   b) $f^{\frac{3}{2}} \times f^{-\frac{1}{3}}$
   c) $g^{\frac{1}{2}} \div g^{2\frac{1}{3}}$
   d) $\frac{h^{\frac{2}{3}} \times h^{\frac{5}{2}}}{h^{\frac{1}{2}}}$

8. Simplify the following:
   a) $(64q)^{\frac{1}{3}}$
   b) $s^{\frac{2}{3}} \div s^{-\frac{2}{3}}$
   c) $d^2 \times \frac{1}{\sqrt{d}} \times \sqrt[4]{d^3}$
   d) $\left(\frac{h^2}{4h^{\frac{1}{2}}}\right)^{\frac{3}{2}}$
   e) $q^{\frac{1}{2}} \times q^{-\frac{2}{3}} \div q^{\frac{3}{4}} \div q^{-\frac{3}{2}}$

Algebra

Variation

Extended

Direct variation

The term **direct variation** is an alternative to **direct proportion**.

Direct variation occurs when there is a direct connection or relationship between two variables. The connection is a multiplier, meaning that the ratio between the two variables is constant. As one variable increases so does the other.

For example:

- 1 mile = 1.6 kilometres   The multiplier connecting miles and kilometres is 1.6
- 20 gallons = 91 litres   The multiplier connecting gallons and miles is $\frac{91}{20} = 4.55$

The symbol for variation, or proportion, is $\propto$

The statement 'cost is proportional to the number of units purchased' is written as: cost $\propto$ units.

The symbol $\propto$ cannot be used to form an equation, but $\propto$ is replaced by ' $= k$', where k is the multiplier that connects the two variables: cost $= k \times$ units.

k is the **constant of proportionality**.

Example

The electricity cost for a household is directly proportional to the number of units of electricity used.
1200 units cost $108

Find **i)** the cost (c) of using 1000 units (u), and **ii)** the number of units used if the cost is $121.50

i) Cost $\propto$ units used $\to c = k \times u \to 108 = k \times 1200 \to \frac{108}{1200} = k \to k = 0.09$

0.09 is the multiplier, or the constant of proportionality, so the formula is: $c = 0.09u$

The cost of 1000 units is: $1000 \times 0.09 = \$90$

ii) When the cost is $121.50 $\to 121.50 = 0.09u \to u = \frac{121.50}{0.09} = 1350$ units used

Direct proportions involving squares, cubes, square roots and cube roots

The method and notation used is the same as before. The extra thing to remember is to apply the power or the root.

Suppose the cost (c) of laying a concrete floor varies directly with the square root of the floor area (a). It costs $396 for 25 m².

What is the cost of laying a concrete floor area measuring 50 m² (to nearest $)? What size of floor area would cost $750 (to 1d.p.)?

Cost $\propto \sqrt{\text{floor area}} \to c = k \times \sqrt{a} \to 396 = k \times \sqrt{25} \to k = \frac{396}{\sqrt{25}} \to k = 79.2$

> **Revision tip**
>
> As with ratio, the order in which the variables are described is the order that you should write them in the statement of proportionality.

79.2 is the multiplier, the constant of proportionality, so the formula is: $c = 79.2\sqrt{a}$

The cost of a floor measuring 50 m² is: $79.2 \times \sqrt{50} = \$560$

When a floor costs \$750 then $750 = 79.2\sqrt{a} \rightarrow \frac{750}{79.2} = \sqrt{a} \rightarrow \left(\frac{750}{79.2}\right)^2 = a \rightarrow a = 89.7$ m²

Inverse variation

The term **inverse variation** is an alternative to **inverse proportion**.

Inverse variation means there is an **inverse** connection between the two variables. One variable is directly proportional to the reciprocal of the other. The product of the two variables is the constant element. As one variable increases, the other decreases.

As your speed (S) over a journey increases, the time (T) taken will decrease. Speed and time have an inverse variation. Speed is inversely proportional to time.

This is written as $S \propto \frac{1}{T}$, and can be rewritten as $S = \frac{k}{T}$, where k is the constant of proportionality.

The time (t) it takes to fill a washbasin with water is indirectly proportional to the rate of flow (f) of water from a tap.

Suppose water is flowing from a tap at 0.2 litres/second and it takes $1\frac{1}{8}$ minutes (m) to fill a wash basin.

How many minutes will it take to fill the wash basin if water is flowing at 0.25 litres/second?

If it takes 2 minutes to fill the wash basin. What is the rate of water flow?

$1\frac{1}{8}$ minutes is 67.5 seconds.

$t = \frac{k}{f} \rightarrow 67.5 = \frac{k}{0.2} \rightarrow k = 67.5 \times 0.2 = 13.5 \rightarrow t = \frac{13.5}{f}$

If water is flowing at 0.25 litres/second, $t = \frac{13.5}{0.25} \rightarrow t = 54$ seconds = 0.9 minutes

If the time is 2 minutes (120 seconds), $120 = \frac{13.5}{f} \rightarrow f = \frac{13.5}{120} \rightarrow f = 0.1125$ litres/second

Quick test

Extended

1. Given that $y \propto x$ and $y = 12$ when $x = 9$, find the value of k and then find:
   a) x when $y = 72$
   b) y when $x = 3$

2. y is directly proportional to the square of x, so that $y = kx^2$. Given that $y = 36$ when $x = 3$, find the value of k and then calculate:
   a) the value of y when $x = 5$
   b) the value of x when $y = 64$

3. y is inversely proportional to x so that $y = \frac{k}{x}$. When $x = 10$ the value of y is 2. Find the value of k and then find:
   a) the value of y when $x = 40$
   b) the value of x when $y = 2.5$

Linear programming

Extended

Graphical inequalities

Linear inequalities can be plotted as lines on a graph. The line forms a boundary between two regions of the graph. The **regions** show where possible solutions to the inequality lie.

For an inequality the statement about the line will include one of the symbols $<$, $\leq$, $\geq$ and $>$, instead of $=$

The lines will be solid or dashed, depending on the inequality sign.

- For a strict inequality ($<$ or $>$) the line is dashed, showing that coordinate points on the line are not included in the solution.
- For an inequality of the type $\leq$ or $\geq$, the line is solid line, indicating that coordinate points on the line are included in the solution.

After drawing the line, you need to shade in the region that **does not** satisfy the inequality. To do this, choose a coordinate point that is not on the line and substitute the x- and y-values into the inequality. If the line does not pass through the origin (0, 0) then use this as the test coordinate. If your chosen point does not satisfy the inequality then it lies in the unwanted region, so shade in that side of the line.

> **Revision tip**
>
> Confusion between the significance of a dashed line and a solid line is a common problem. Make sure you remember what each type of line means.

For example:

a) $x \geq 2$

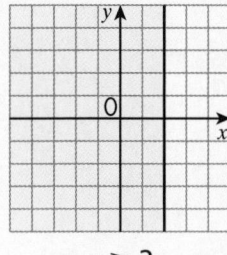

b) $y < -x$

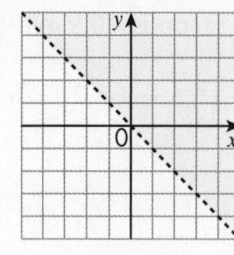

c) $-1 < x \leq 4$

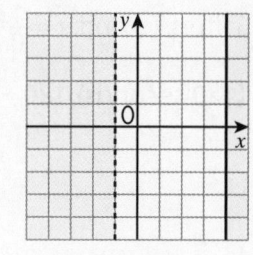

d) $4x + 1 \leq y$

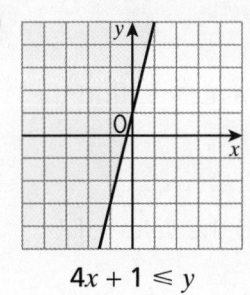

e) $4x + 6y > 12$

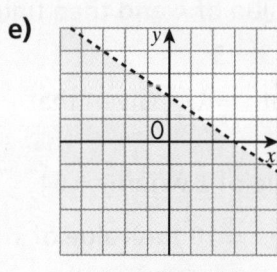

More than one inequality

When you need to draw more than one inequality on the same set of axes, you are being asked to define a **region**, often a triangle or quadrilateral. In this case it is usual to shade on the side of the inequality that is not needed, leaving the required region blank.

Example

a) On the same grid identify the region defined by these inequalities:
$x \geqslant 1$, $y > \frac{1}{2}x + 2$, $y < 8 - x$

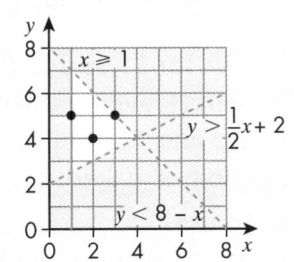

b) Are the points (3, 5), (2, 4), (1, 5) in the defined region?

(3, 5) ✗ on the dashed line; (2, 4) ✓ inside the region; (1, 5) ✓ on the solid line

c) What is the largest integer y-coordinate of any point in the region?

6

d) What is the largest integer x-coordinate of any point in the region?

3

e) What is the largest integer value of $x + y$ in the region?

7 (1, 6) and (3, 4)

Linear programming

When solving practical problems, a solution may not always be a simply defined region and there may be multiple solutions.

Plotting multiple inequalities on a graph enables you to solve practical problems.

Example

A hotel is having an extension built. The floor area of the extension can be a maximum of 128 m². Single rooms will have an area of 8 m² (number of single rooms = x) and double rooms will have an area of 16 m² (number of double rooms = y). There must be at least 10 rooms in total and there must be fewer than 7 single rooms.

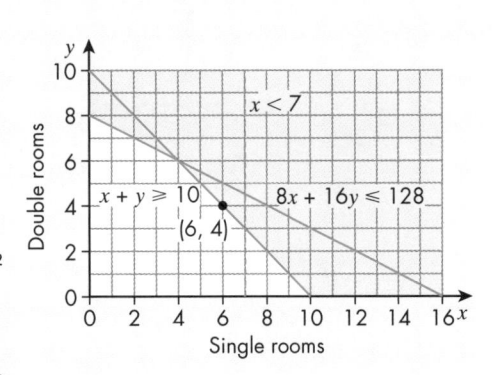

a) What are the three inequalities that the hotelier needs to use?

$x + y \geqslant 10$, $8x + 16y \leqslant 128$, $x < 7$

b) Plot these inequalities on a graph and identify the region where solutions can be found.

c) What is the maximum number of single rooms that will fulfil the hotelier's needs?

6 single rooms (and 4 double rooms)

Algebra

Quick test

Extended

1. a) On the same grid identify the region defined by these inequalities.
       $y \leq 6$, $x > -2$, $4x + 2y \leq 12$, $2x - 3y > 6$
    b) What is the greatest integer value of y in the region?
    c) What is the greatest integer value of x in the region?
    d) Which of these coordinate points is satisfied by all four inequalities?
       (1, 4)   (1, −1)   (−2, 6)   (−1, −2)

2. A baker makes two different size loaves. The small loaf sells for $1.50 and the large for $3.00. His oven will take a maximum of 30 baking tins and he only puts loaves in the oven when the sale price of the batch is at least $60. The baker has 18 large loaf tins and 14 small loaf tins. If x represents the number of small loaves baked and y the number of large loaves baked:
    a) What are the four inequalities that the baker needs to use?
    b) Plot these inequalities on a graph and identify the region where solutions can be found.
    c) What combination of loaves x and y gives the greatest batch sale price?

Functions

Extended

Function notation

When you write $y = 4x + 2$ or $y = 3x^2 + 4x - 3$, the equations are showing that y is a **function** of x; the value of y depends on the value of x.

You could write the first equation above as $f(x) = 4x + 2$, then you would call it 'function f'.

The value of the function when $x = 7$ is written as: $f(7)$, so $f(7) = 4 \times 7 + 2 = 30$

Instead of $f(x) = 4x + 2$ you could write f: $x \to 4x + 2$. This is just another way of writing a function. They mean the same thing.

If there are several equations in the same problem, it is useful to call them functions, with different identifying letters:

$f(x) = 4x + 2$, $g(x) = 3x^2 + 4x - 3$

Inverse functions

The **inverse function** has the opposite, or reverse, effect of the original function.

In $f(x) = 4x + 2$ the function 'multiplies x by 4 and adds 2'

The reverse will be 'subtract 2 and then divide by 4', so $f^{-1}(x) = \frac{x-2}{4}$
Then $f^{-1}(10) = 2$ and $f^{-1}(4) = 0.5$

> **Revision tip**
>
> If $f(n)$, then $f^{-1}f(n) = n$, which allows you to check that you have created the inverse correctly.

To find the inverse of a function:

Write $y = f(x)$

$y = 4x + 2$

Then rearrange to make x the subject: $y - 2 = 4x \to x = \frac{y-2}{4}$

Finally, replace y in the answer with x: $f^{-1}(x) = \frac{x-2}{4}$

Composite functions

Suppose $m(x) = 3x$ and $n(x) = x + 2$ and the result for $m(x)$ is input to $n(x)$.

$m(5) = 3 \times 5 = 15 \to n(15) = 15 + 2 = 17$

This can be written as a **composite** function: $nm(x) \to nm(5) = 17$

In the same way, $mn(4) = 18$ since $n(4) = 6$ and then $m(6) = 18$

Notice that the right-most function is carried out first and then the next function to the left.

If $p(x) = x^2$ and $q(x) = 6 - x$

then $pq(8) = 4$ since $q(8) = -2$ and then $p(-2) = 4$

and $qp(3) = -3$ since $p(3) = 9$ and then $q(9) = -3$

More about composite functions

If you have two functions, $c(x) = x^2 - 3$ and $d(x) = 3x + 1$, it is possible to combine these to give a single expression for cd(x)

First of all evaluate the two functions as described above with a value of, say, $x = 2$

cd(2) = 46 since d(2) = 7 and then c(7) = 46

cd(x) means, 'start with x, multiply it by 3 and add 1' which gives $3x + 1$, then 'take that answer, square it and subtract 3', which is $(3x + 1)^2 - 3$.

Now simplify: $(3x + 1)^2 - 3 \rightarrow (9x^2 + 6x + 1) - 3 \rightarrow 9x^2 + 6x - 2$

So cd(x) = $9x^2 + 6x - 2$

To check, try cd(x) = $9x^2 + 6x - 2$ with a value of $x = 2$

cd(2) = $9(2)^2 + 6(2) - 2 \rightarrow 36 + 12 - 2 \rightarrow 46$

In the same way finding a single expression for dc(x) would produce dc(x) = $3x^2 - 8$.

> **Revision tip**
>
> Remember that gf means 'f first, then g' when functions are combined.

Quick test

Extended

1. $f(x) = 2x^2 - 4$

   Find the value of these functions.

   a) f(2)　　　b) f(6)　　　c) f(−3)　　　d) f(½)

2. a) Find $f^{-1}(x)$ for each of these functions.

   i) $f(x) = 2x - 7$　　ii) $f(x) = \frac{x}{3} + 4$　　iii) $f(x) = \frac{6}{x+1}$　　iv) $f(x) = \frac{2x-3}{5}$

   b) For each of the inverse functions find $f^{-1}(4)$

3. If $p(x) = 4x + 2$ and $q(x) = x^2 - 3$, find:

   a) p(3)　　　b) q(4)　　　c) pq(−2)

4. If $s(x) = 5x - 3$ and $t(x) = \sqrt{x}$, find:

   a) st(16)　　b) ts(5.6)　　c) ss($\frac{1}{2}$)　　d) tt(81)

5. Find a single expression for ts(x) when $s(x) = 5x - 3$ and $t(x) = \sqrt{x}$

Exam-style practice questions

1 **a)** Maddie went clothes shopping. She bought t tops priced at \$15, j jeggings priced at \$20 and s pairs of sports socks priced at \$1.50 [2]

Write an equation to find the cost (c) of Maddie's clothes in terms of t, j and s.

b) A parcel delivery service uses this rule to calculate the cost of delivering a parcel:

Cost = \$2.70 plus \$1.25 per kilogram

   i) How much does it cost to send a 5 kg parcel? [1]

   ii) Jamilla pays \$13.95 to send a parcel. What was its mass? [3]

c) $c = s^2 + 1.25k$

   i) Make k the subject. [2]   **ii)** Make s the subject. [2]

d) $v = \frac{1}{3}\pi r^2 h$

   i) Make h the subject. [2]   **ii)** Make r the subject. [3]

2 **a)** Simplify these expressions. **i)** $3p \times 4q$ [1]   **ii)** $10m^2 \times \frac{1}{2}m^2$ [2]

b) A parallelogram has side lengths of $(2x + 3)$ and $(16 - x)$. Write an expression, in its simplest form, for the perimeter. [2]

c) Expand these expressions. **i)** $6(v + 2w)$ [2]   **ii)** $4m(2p^2 - 3m)$ [2]

d) Expand and simplify these expressions.

   i) $2(m + 3) + 4(2m - 3)$ [3]   **ii)** $3p(2q + p^2) - 5p^2(q - 2p)$ [3]

e) Factorise these expressions. **i)** $6v - 15w$ [2]   **ii)** $6ab^2c + 15b^3c^3$ [2]

f) Expand these expressions. **i)** $(d + 2)(d - 5)$ [2]   **ii)** $(5 - e)(6 + e)$ [2]

g) Expand these expressions. **i)** $(2g + 5h)(4g - 3h)$ [2]   **ii)** $(3f - 2)(3 - 4f)$ [2]

h) Expand these expressions. **i)** $(5m + 2n)(5m - 2n)$ [2]   **ii)** $(pq + st)(pq - st)$ [2]

i) Expand these expressions. **i)** $(6a - 5)^2$ [2]   **ii)** $(b - 3)^2 - 1$ [2]

Extended

j) Factorise these expressions. **i)** $c^2 - 2c - 8$ [2]   **ii)** $f^2 - 5f + 6$ [2]

k) Factorise these expressions. **i)** $b^2 - 49$ [2]   **ii)** $25e^2 - 100g^2$ [2]

l) Factorise these expressions. **i)** $6x^2 + 3x - 3$ [3]   **ii)** $5x^2 - 13x - 6$ [3]

m) Simplify these expressions. **i)** $\frac{2}{x+3} + \frac{1}{2x-3}$ [3]   **ii)** $\frac{x^2 - 2x - 3}{2x^2 - 2x - 12}$ [3]

n) A cereal packet is in the shape of a cuboid. It has side lengths $(x + 3)$ cm, $(2x - 1)$ cm and $(x + 2)$ cm. Find a simplified expression for the volume of the cuboid. [2]

3

a) Solve these equations.   i) $\frac{4d+7}{3} = 9$ [2]   ii) $\frac{20-7c}{3} = 2$ [2]

b) Solve these equations.   i) $2(m+4) = 14$ [2]   ii) $3(3p+10) = 12$ [2]

c) Solve these equations.   i) $2q + 3 = 3q - 2$ [2]   ii) $4s + 5 = 2s - 1$ [2]

d) Two coaches, A and B, are used to take 92 students on a trip. To start with there are 10 more students on coach A than on coach B. Just before departure 7 students move from coach A to coach B to be with friends. How many students were on each coach when they set off?   [2]

Extended

e) Solve these equations.   i) $x^2 + 3x - 28 = 0$ [2]   ii) $x^2 - 7x = -10$ [2]

f) Solve these equations.   i) $6x^2 - 21x = 0$ [2]   ii) $9x^2 + 7x + 4 = 6 + 2x - 3x^2$ [3]

g) A rectangular bathroom wall is 6 m wider than it is high. The area of the wall is 21.25 m².

   i) What are the dimensions of the wall?   [2]

   ii) The wall is going to be covered with square tiles, each with an area of 225 cm². Can the wall be covered without cutting any tiles?   [2]

h) i) Rewrite the expression $x^2 + 9x - 6$, in the form $(x \pm a)^2 - b$.   [2]

   ii) Solve $3x^2 - 12x - 30 = 0$ by completing the square, leaving the answer in surd form.   [3]

i) Solve the simultaneous equations. $3x + y = 11$
   $2x - y = 9$ [4]

j) Solve the simultaneous equations. $2x + 8y = -2$
   $x + 2y = -2$ [4]

k) Solve the simultaneous equations: $4x + 3y = -3$
   $8x - 4y = 14$ [4]

Extended

l) Ruby and Sally are sisters. Their combined ages are 16 years. In 4 years time, Sally will be twice as old as Ruby. How old are they now?   [2]

m) i) Find the smallest positive integer that will satisfy the linear inequality $20 - 3x \leq 3$ [1]

   ii) Solve the inequality $\frac{x-6}{5} \leq 3$ [2]

n) Solve these pairs of simultaneous equations:

$x^2 + y^2 = 20$ and $y = 3x - 10$ [4]

o) An equilateral triangle has a side length of $(72 - 4x)$ cm. The perimeter of the triangle is less than 84 cm. Show that x can be represented by the diagram:

```
   o─────────────o
11  12  13  14  15  16  17  18
```
[2]

4 a) The table shows some conversions between litres and gallons.

Gallons	4	8	12
Litres	18	36	54

    i) Draw a graph to show this information. [3]

    ii) How many litres would a car petrol tank of capacity 15 gallons hold? [1]

b) A train travels at a steady speed of 120 km/h for 1.5 hours. It waits at a station for 15 minutes and then completes the remaining 60 km its journey in 45 minutes.

    i) Show the journey of the train as a distance–time graph. [4]

    ii) On the graph, draw a line representing the average speed for the journey. Find this average speed. [1]

Extended

c) This **velocity–time** graph shows the speed of an aircraft coming onto the radar of an air-traffic control tower on its approach to, and eventual landing at, an airport.

What was the total distance travelled over the 60 minutes? [3]

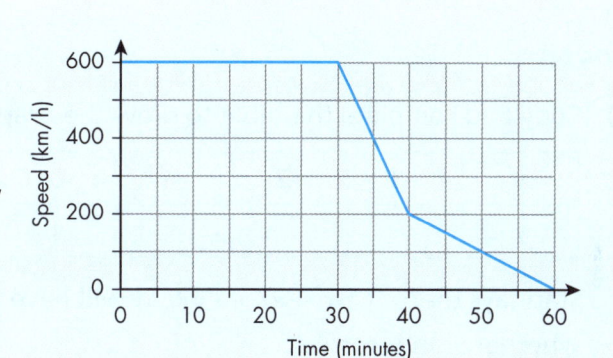

5 a) On the same axes, draw the graphs of $y = 2x - 2$ and $y = \frac{1}{2}x + 4$ for $0 \leq x \leq 6$.

At what point do the two lines intersect? [4]

b) The line $y = mx - 2$ passes through the point (4, 10).

What is the value of m, the gradient of the line? [2]

c) A line with the equation $y = mx + 1$ passes through the point (2, 7). Another line parallel to this one passes through (3, 6). What is the equation of the second line? [3]

Extended

d) A triangle has vertices with coordinates $A(4, 8)$, $B(2, 2)$ and $C(8, 4)$.

    i) Find the equations of the sides AB and AC. [4]

    ii) Find the midpoint of BC. [1]

e) A line segment has end points $A(2, 3)$ and $B(6, 5)$. Find the equation of the perpendicular line that passes through the midpoint of AB. [4]

6 **a)** Copy and complete the table to draw the graph for the equation $y = x^2 + 2x - 3$.

x	−4	−3	−2	−1	0	1	2
y	5				0		

  i) Use the graph to solve the equation $3 = x^2 + 2x - 3$. [6]

  ii) What are the coordinates of the lowest point on the graph? [1]

b) Copy and complete the table to draw the graph for equation $y = \dfrac{4}{x}$.

x	0.5	1	2	4	8	16
y	8					0.25

On the same axes draw the line for $y = 0.5x - 1$ then use the graph to solve the equation $\dfrac{1}{x} = 0.5x - 1$. [7]

Extended

c) Copy and complete the table to draw the graph for the equation $y = x^3 + 3x^2 + 2x + 1$

x	−3	−2	−1	0	1
y				1	

Stan says that $x^3 + 3x^2 + 2x + 1 = 0.5x$ will have three solutions. Use your graph to decide whether Stan is correct. [7]

d) In the 2016 census, an island has a population of 25 000. The net migration (population leaving) per year is measured at 10% per annum.

Copy and complete (to 2 s.f.) the table to draw the graph for the decline in population.

Year	2016	2018	2020	2022	2024	2026
Population	25 000					8700

Estimate the year when the population will first drop below 12 000. [5]

e) Look at the graph drawn in **Q6c)** for $y = x^3 + 3x^2 + 2x + 1$.

Estimate the gradients at (−2, 1) and (0, 1). [4]

7 **a)** Look at these sequences. Find the term-to-term rule for each sequence and find the next two terms.

  i) 3, 7, 11, 15, … [2]   **ii)** 200, 40, 8, 1.6, … [2]

b) Write down the first four terms of the sequence with nth term:

  i) $2n + 3$ [2]   **ii)** $n^2 - 3$ [2]   **iii)** $\dfrac{n^3}{5}$ [2]

c) Work out the nth term for each of these sequences:

  i) 2, 6, 10, 14, … [2]   **ii)** 4, 7, 12, 19, … [2]   **iii)** 2, 16, 54, 128, [2]

d) A pattern of squares is built up as shown in the diagram.

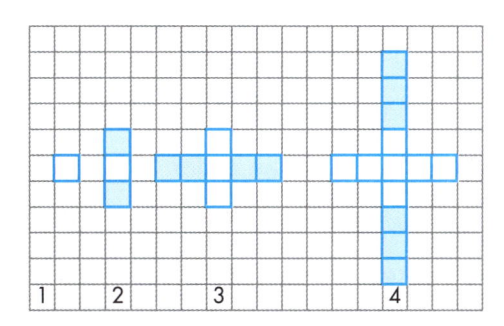

i) How many squares will be in the 5th and 6th patterns? [2]

For the nth pattern the number of squares is $n^2 - n + 1$

ii) Show that this formula gives the correct answer for $n = 7$ [1]

iii) How many squares will there be in the 100th pattern? [1]

Extended

e) The first five terms of a sequence are: 1, 10, 23, 40, 61, …

i) Find the nth term of the sequence. [2]

ii) What is the 100th term in this sequence? [1]

iii) Are the terms 5351 and 3361 in the sequence? Give reasons for your answers. [2, 2]

f) A sequence has the first 5 terms: 3, 24, 81, 192, 375

i) Find the nth term of this sequence. [2]

ii) Find x_{10} and x_{25} [2]

8 a) Work out the value of each of these power terms.

i) 4^5 [1]   ii) 2.7^4 [1]   iii) 5.1^0 [1]   iv) 7.2^1 [1]

b) i) Write 7^{-3} as a fraction. [1]   ii) Write $6b^{-2}$ as a fraction. [1]

iii) Write $\dfrac{5}{c^4}$ in index form. [1]

c) Simplify these expressions: i) $6f^2 \times 3f^3$ [2]   ii) $(3v^3)^2$ [2]   iii) $\dfrac{12m^3n^4}{4m^2n^5}$ [2]

d) Evaluate each number.   i) $121^{\frac{1}{2}}$ [1]   ii) $4913^{\frac{1}{3}}$ [1]   iii) $\left(\dfrac{121}{169}\right)^{\frac{1}{2}}$ [1]   iv) $49^{-\frac{1}{2}}$ [1]

Extended

e) Write in index form:

i) $\dfrac{1}{w^{\frac{3}{4}}}$ [2]   ii) $\sqrt[5]{x^3}$ [2]

Simplify these expressions:

iii) $y^{\frac{3}{4}} \times y^{-\frac{1}{2}} \div \sqrt[3]{y}$ [2]   iv) $(27z^2)^{\frac{1}{3}}$ [2]

Solve this equation:

v) $x^{-\frac{1}{3}} = 2x^{-1}$ [2]

Extended

9 **a)** A potter can make 25 statuettes from 3.5 kg of clay. The number of pottery statuettes that she can make is proportional to the mass of clay available.

   i) What mass of clay is needed to make an order of 60 statues? [2]

   ii) How many complete statues can be made from 5 kg of clay? [1]

   b) A jewelled pendant is made in several sizes. The mass is directly proportional to the cube of its length. A pendant 2 cm long has a mass of 36 g.

   i) What is the mass of a pendant of length 4 cm? [2]

   ii) What is the length of a pendant that has a mass of 121.5 g? [1]

   c) The number of shares in a company held by each investor is inversely proportional to the number of investors in the company.

   When there are 200 investors they each have 1250 shares.

   i) How many shares would be held by each investor if there were 625 investors? [2]

   ii) Each investor received 1000 shares. How many investors were there? [1]

10 **a)** On a set of axes draw the line $y = 2 - 0.4x$ as a dashed line.

   Shade in the region defined by $y > 2 - 0.4x$ [4]

   b) On the same grid draw the region, A, defined by the inequalities $y < 2x + 3$, $y > 0.5x + 1$, $2y + 3x \leq 9$

   Shade the region that is not required. [5]

   How many points (with integer coordinates) in the region are defined by these inequalities? [1]

   c) A gardener has $120 to spend on new shrubs: x small shrubs, each costing $8, and y large shrubs, each costing $15. She does not want to buy more than 10 in total. She needs more than 2 small shrubs and more than 4 large shrubs.

   i) Write the four equalities, in terms of x and y, that the gardener will need to use. [4]

   ii) Show on a graph the region where these four inequalities are satisfied. [4]

   iii) List the possible combinations of small and large shrubs that she can buy. [2]

11 **a)** $m(x) = 2x^2 - 3$ and $n(x) = 6x - 3$

   i) Find the value of $m(5)$ and $n\left(\frac{1}{2}\right)$ [1, 1]

   ii) If $m(a) = 29$, what are the values of a? [2]

   iii) What are the values x for which $m(x) = n(x)$? [2]

   b) $p(x) = \frac{3}{2x - 1}$

   i) Find an expression for $p^{-1}(x)$ [3]   **ii)** Find the value of $p^{-1}(6)$ [1]

   c) $m(x) = 2x^2 - 3$ and $n(x) = 6x - 3$

   i) Find the value of $mn(3)$ [2]   **ii)** Find the value of $nm(2)$ [2]

   d) $m(x) = 2x^2 - 3$ and $n(x) = 6x - 3$

   i) Find an expression for $mn(x)$ [2]   **ii)** Find an expression for $nm(x)$ [1]

Angle properties

Angle facts

The **angles on a straight line** add up to 180°

$a + b = 180°$

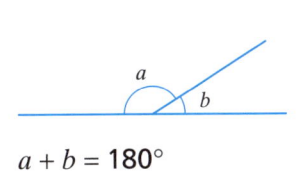

$55° + x + 20° = 180°$
$x = 180° - 55° - 20°$
$x = 105°$

The sum of the **angles around a point** is 360°

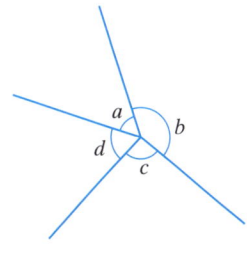

$a + b + c + d = 360°$

$x + 40° + 115° + 100° = 360°$
$x = 360° - 40° - 115° - 100°$
$x = 105°$

Opposite angles, also called **vertically opposite angles**, are equal.

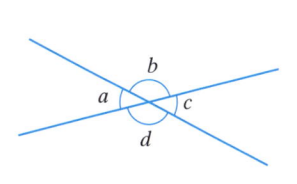

$a = c$ and $b = d$

$a = 130°$
$b = 180° - 130° = 50°$
$c = 50°$

Angles in parallel lines

Parallel lines are lines that are always the same distance from each other. They are identified by small arrowheads drawn on them. A line cutting across a pair of parallel lines is called a transversal.

Geometry 87

Corresponding angles are equal. (Look for the letter F.)	**Alternate angles** are equal. (Look for the letter Z.)	**Interior angles** (sometimes called **allied angles**) add up to 180°. (Look for the letter C.)
F	Z	C

When you are asked to provide reasoning in your solution, use the words 'corresponding', 'alternate' and 'interior'. You should not refer to 'F, Z or C shapes'.

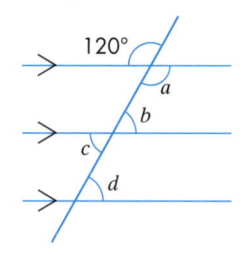

$a = 120°$ (vertically opposite to 120°)
$b = 60°$ (interior angle with a)
$c = 60°$ (vertically opposite to b)
$d = 120°$ (corresponding to b) or (alternate to c)

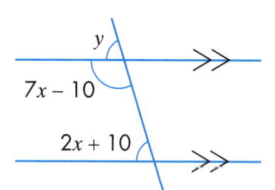

$(7x - 10°) + (2x + 10°) = 180°$ (interior angles)
$9x = 180°$
$x = 20°$
$y = 2(20°) + 10°$ (corresponding angles are equal)
$y = 50°$

> **Revision tip**
>
> Questions asking you to work out the size of an angle will usually ask you to provide a reason to support your answer so it is important to learn these short phrases.

Angles in a triangle

In any triangle, the three angles add up to 180°.

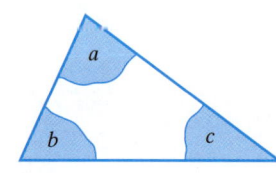

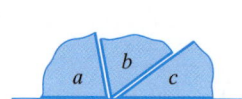

The three angles in a triangle = three angles on a straight line.

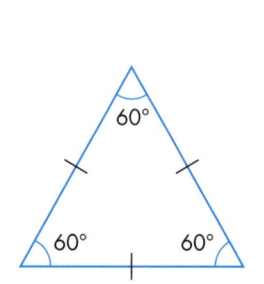

Equilateral triangles:
three sides are equal
three angles are equal

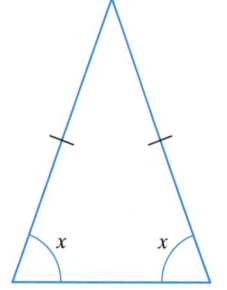

Isosceles triangles:
two sides are equal
two base angles are equal

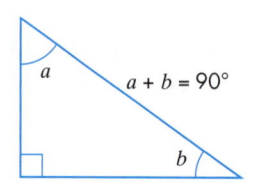

Right-angled triangles:
have one angle of 90°, so the other two angles add up to 90°

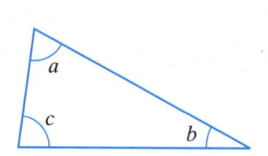

Scalene triangles:
the sides and angles of a scalene triangle are all different

88 IGCSE Mathematics Revision Guide

In the triangle below, $a = 180° - 150°$ (angles on a straight line add to 180°)

$a = 30°$

$x + x + 30° = 180°$ (angles in a triangle add up to 180° and base angles of an isosceles triangle are equal)

$x = 75°$

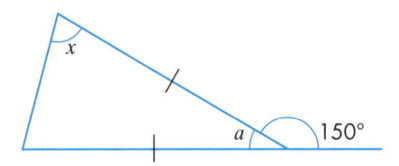

Angles in a quadrilateral

The four angles of a **quadrilateral** add up to 360°.

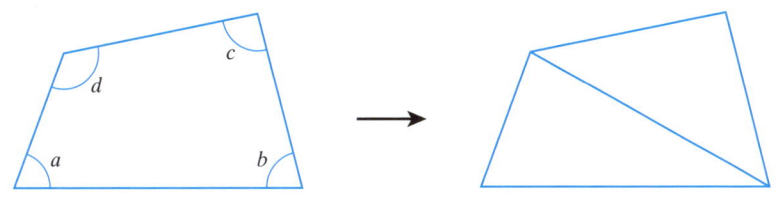

$a + b + c + d = 360°$ because any quadrilateral can be cut into two triangles, and $2 \times 180° = 360°$

Special quadrilaterals

Rhombus	Parallelogram
• Two pairs of parallel sides • Four equal sides • Opposite angles are equal • Diagonals **bisect** each other at 90° • Diagonals bisect each of the angles at the four vertices	• Two pairs of parallel sides • Opposite sides are equal • Opposite angles are equal • Diagonals bisect each other
Kite	**Trapezium**
• Two pairs of equal adjacent sides • Opposite angles between the sides of different lengths are equal • Diagonals intersect at 90°	• One pair of parallel sides • The angles at the ends of the line running across the parallel lines are interior angles and add up to 180°

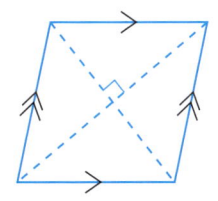

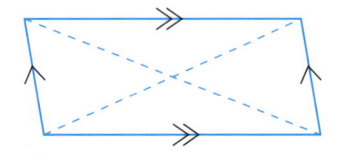

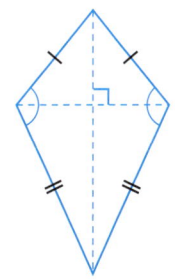

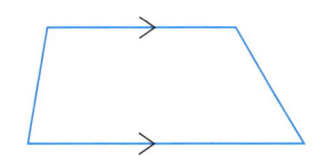

> **Revision tip**
>
> Squares and rectangles are also special quadrilaterals – make sure you know their properties.

Regular and irregular polygons

A **polygon** is a closed shape with three or more straight sides.

A polygon is regular if all its **interior angles** are equal and all its sides are the same length.

Any other polygon is irregular.

Here are five regular polygons.

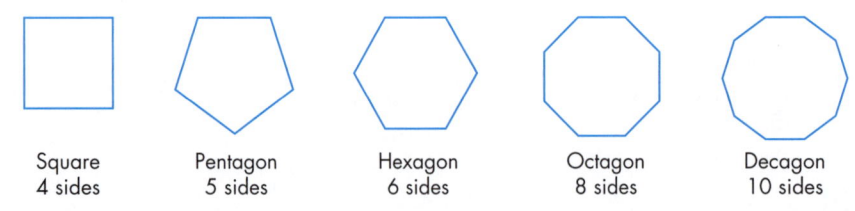

Square	Pentagon	Hexagon	Octagon	Decagon
4 sides	5 sides	6 sides	8 sides	10 sides

As well as these, a heptagon has 7 sides and a nonagon has 9 sides.

The sum, S, of the interior angles for an n-sided polygon is given by the formula: $S = 180(n - 2)°$

This is because any polygon can be split into a number of triangles from one starting **vertex**.

The number of triangles is always $n - 2$, and the sum of the angles in each triangle is 180°

If the polygon is regular, each interior angle is $\frac{(n-2) \times 180°}{n}$

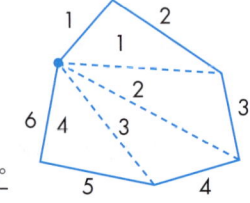

For any polygon, the sum of the **exterior angles** is always 360°, so each external angle of a regular polygon (marked as y in the diagram) is $\frac{360°}{n}$

In any polygon, the interior and exterior angles at a vertex add up to 180°, because they are angles on a straight line and they sum to 180°

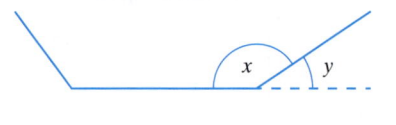

$x + y = 180°$

A vertex of a regular polygon has an internal angle of $4x$ and an external angle of $(x + 5°)$. What sort of polygon is it?

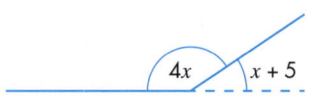

$4x + (x + 5°) = 180°$ (angles on a straight line = 180°)

$5x = 175° \rightarrow x = 35° \rightarrow$ the external angle is 40°

The number of sides is $\frac{360°}{40°} = 9$ The polygon is a nonagon.

Tangents and diameters

A tangent is a straight line that touches a circle at a single point. If a radius is drawn to meet the point of contact it forms a right angle with the tangent – the radius is perpendicular to the tangent.

Two tangents drawn from a common point are of equal length.

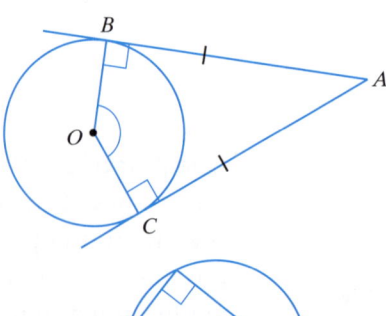

O is the centre of the circle.
$AB = AC$
Angle ABO = Angle ACO = 90°

> **Revision tip**
>
> If a connection between angles in various parts of a diagram is not obvious, rotate the page and look at the diagram from different angles. A different view often helps.

Triangles drawn on a diameter always have a right angle at the circumference.

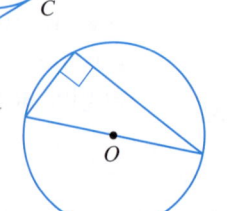

Extended

Angles in a circle
The angle **subtended** by an arc (or chord) at the centre of a circle is twice the angle subtended at the circumference.

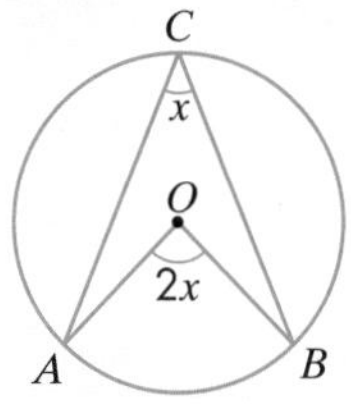

The angles subtended at the circumference by an arc (or chord) are equal.

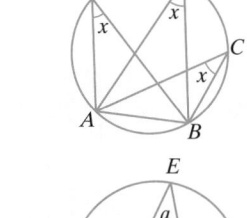

Cyclic quadrilaterals
A **cyclic quadrilateral** is one in which all four vertices lie on the circumference of a circle.

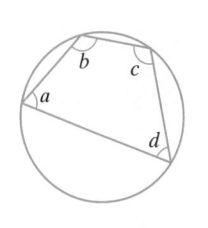

The opposite angles of a cyclic quadrilateral add up to 180°
$a + c = 180°$ and $b + d = 180°$

Alternate segments
A chord cuts a circle into two segments. The angle between a tangent and the chord is equal to the angle subtended from the ends of the chord in the alternate segment.

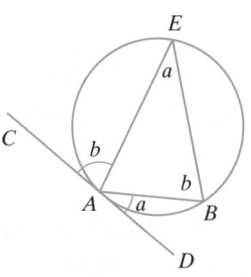

Example
O is the centre of the circle.

DCE is a tangent to the circle at C.

Work out Angle FCE.

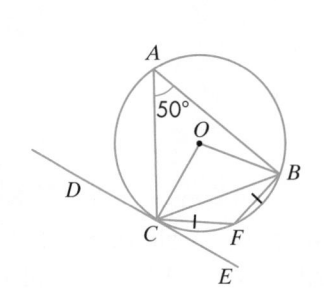

Angle $COB = 100°$ (angle at centre 2 × angle at circumference)
Angle $OCB = 40°$ (triangle OCB is isosceles, $OC = OB$ = radius)
Angle $CFB = 130°$ (opposite angles of cyclic quadrilateral add up to 180°)
Angle $CBF = 25°$ (triangle BCF is isosceles)
Angle $FCE = 90° - 40° - 25° = 25°$ (Angle $OCE = 90°$, radius meets tangent at 90°)

Note that Angle $ECB = 50°$ (equal to Angle CAB because of alternate segment theory), which is also true since Angle CBE = Angle FCE + Angle $FCB = 25° + 25° = 50°$

Quick test

1. In these questions, find the size of each angle marked with a letter.
   a) 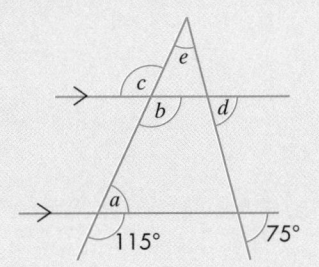
   b)

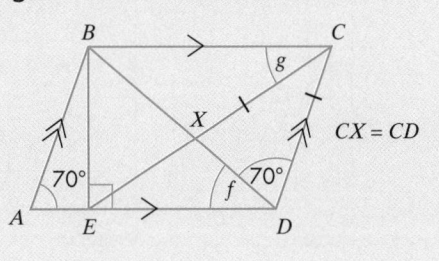

Extended

2. In these questions, find the size of each angle marked with a letter. In each case, O is the centre of the circle.
   a)
   b)

3. A regular polygon has exterior angles of 15°.
   How many sides does it have?

4. A regular polygon has interior angles of 168°.
   How many sides does it have?

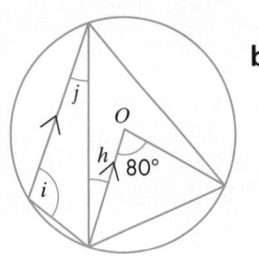

Geometry

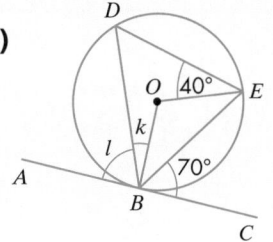

Geometrical terms and relationships

Measuring and drawing angles

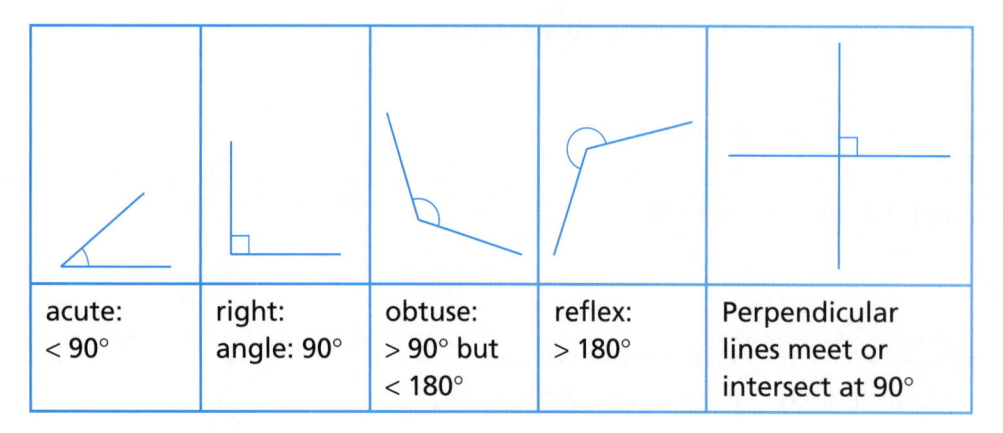

acute: < 90°	right: angle: 90°	obtuse: > 90° but < 180°	reflex: > 180°	Perpendicular lines meet or intersect at 90°

A ski piste rated as a grade 'blue' (intermediate) should not have a slope of more than 18°. Here is a diagram of a skier coming down a piste. Is this a 'blue' piste?

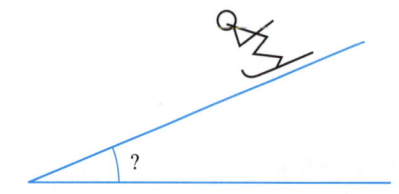

Measure the angle marked '?' with a protractor. You should find that it is 22° → 22° > 18° so the piste is not rated as 'blue'.

Bearings

An angle bearing represents a degree of turn or change of direction. A bearing is measured clockwise from the North, i.e. 000°–360°. A bearing is usually written using three figures (known as a **three-figure bearing**).

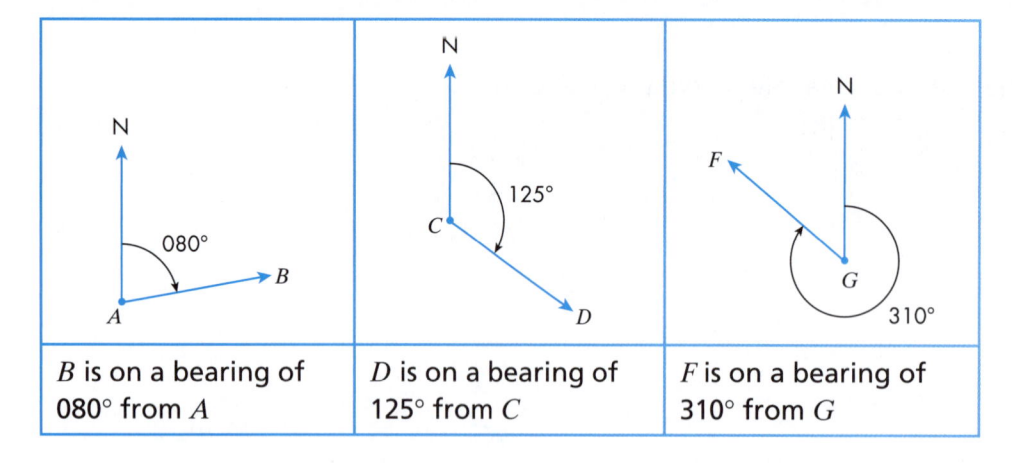

B is on a bearing of 080° from A	D is on a bearing of 125° from C	F is on a bearing of 310° from G

Example

A and B are lights at either side of the entrance to a harbour. A boat (X) is approaching the harbour on a bearing of 050° from A and a bearing of 330° from B. Mark the position of the boat.

Draw lines representing north (000°). Measure and draw a bearing of 50° from A. Measure and draw a bearing of 330° from B.

Where the two lines cross marks the position of the boat.

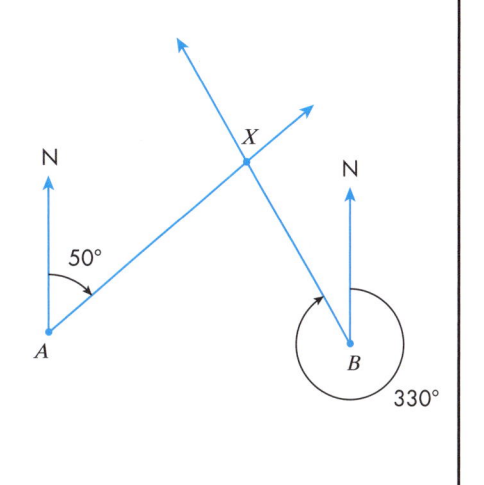

> **Revision tip**
>
> Bearings themselves only indicate direction. If you need to include distance, you will be given a scale.

Nets

A net is a flat shape that can be cut out and folded to make a solid three-dimensional object.

You should know these facts about solid shapes.

- A cube has six square faces.

- A cuboid has six rectangular faces.

- A prism has a uniform **cross-section**.

    Triangular prism   Hexagonal prism

- A **pyramid** has a polygonal base and triangular faces that meet at the vertex.

    Each of these solid shapes can be made from a **net**.

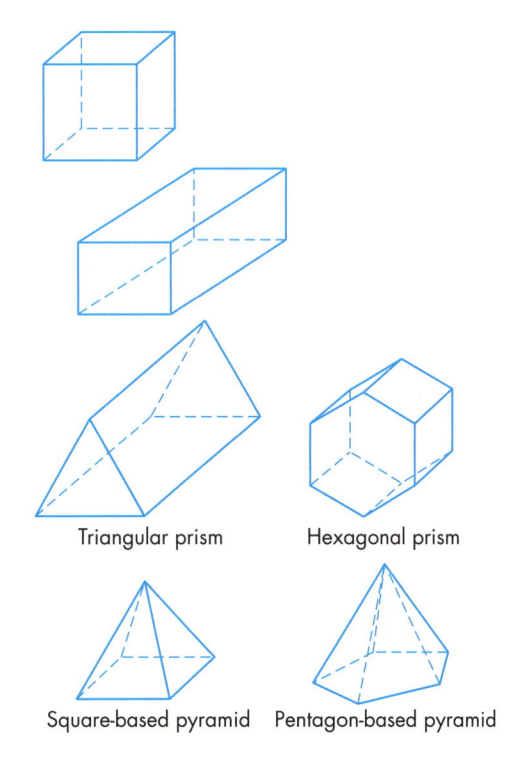

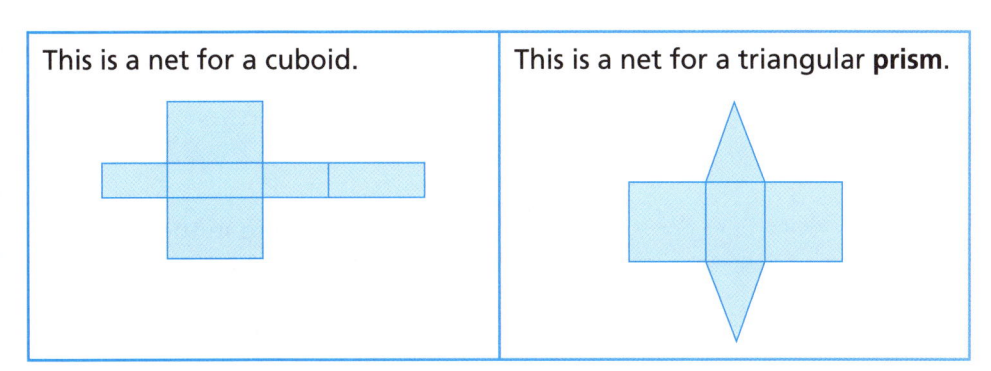

    Square-based pyramid   Pentagon-based pyramid

This is a net for a cuboid.	This is a net for a triangular **prism**.

Geometry 93

Congruent shapes

Any two two-dimensional shapes that are exactly the same size and shape are **congruent** – they are identical in every way. If you cut out, or trace over one of the shapes its outline would fit exactly over the other, even though you might need to turn one shape over.

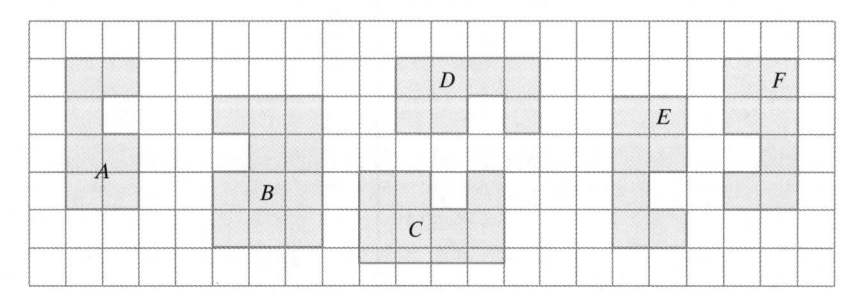

Shapes A, D, E and F are congruent. D and F have been rotated and E is a reflection.

> **Revision tip**
>
> When an object is transformed using a translation, rotation or reflection, the image and object are congruent to each other.

Extended

Congruent triangles

Two triangles are congruent if they satisfy any of the following four rules, or 'tests', of congruency.

Side, Side, Side (SSS)	All three sides of one triangle are exactly the same as the sides of a second triangle
Side, Angle, Side (SAS)	Two sides and the angle between them in one triangle are exactly the same as two sides and the angle between them in the other triangle
Angle, Side, Angle (ASA)	Two angles and one side in one triangle are exactly the same as two angles and the corresponding side in the second triangle
Right angle, Hypotenuse, Side (RHS)	Both triangles are right angled and the hypotenuse and one other side are equal

For example: in each case, explain how you can tell whether the pairs of triangles are congruent or not.

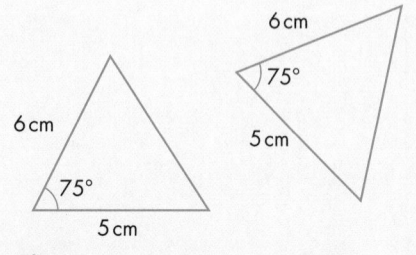

These are congruent – SAS

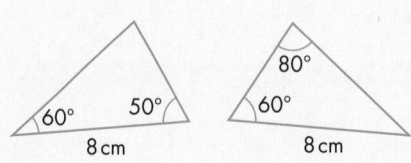

The missing angle in the right-hand triangle is 40°. Because the angles in the two triangles do not match they cannot be congruent.

IGCSE Mathematics Revision Guide

Similar shapes

Two shapes are **similar** if one is an **enlargement** of the other. The angles in the shape remain the same, but the size changes.

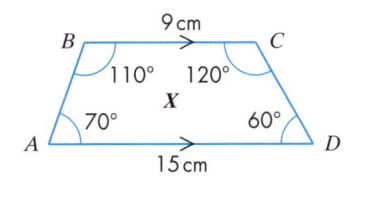

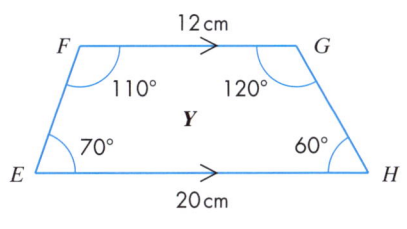

> **Revision tip**
>
> When an object is enlarged the image and object are similar to each other.

The ratio of the lengths of the sides is $15 : 20 = 3 : 4$

The scale factor of enlargement is $\frac{4}{3}$

You need to identify the sides that correspond or match – $AD : EH$, $BC : FE$ and so on.

You can express the change in size as a ratio or as a **linear scale factor**.

The length of AB is measured as 6.4 cm. What is the length of EF?

The factor of enlargement is $\frac{4}{3}$, so $EF = \frac{4}{3} \times 6.4 = 8.53$ cm

The distance between the parallel lines in shape Y is 8 cm. What is the corresponding measurement in shape X?

The factor of enlargement is $\frac{4}{3}$, so $8 = \frac{4}{3} \times$ distance in $X \rightarrow$ distance in $X = \frac{8 \times 3}{4} = 6$ cm

Extended

Areas of similar triangles

Similar triangles have three matching angles. If the linear scale factor is k then the **area scale factor** is k^2.

Triangle ABC is similar to triangle DFE.

So, triangle DFE is an enlargement of triangle ABC with a scale factor of $\frac{DE}{AC} = \frac{EF}{CB} = \frac{FD}{BA} = \frac{10}{4} = 2.5$

The area scale factor is $2.5^2 = 6.25$

What is the area of triangle DFE?

area of $ABC = 6$ cm², area of $DFE = 6 \times 6.25 = 37.5$ cm²

Check: $EF = 3 \times 2.5 = 7.5$

area of $DFE = \frac{1}{2} \times 10 \times 7.5 = 37.5$ cm²

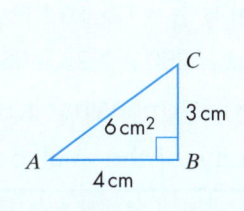

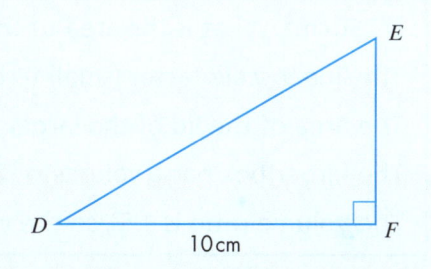

Geometry

Extended (Continued)

> **Example**
>
> Triangles PQR and XYZ are similar. 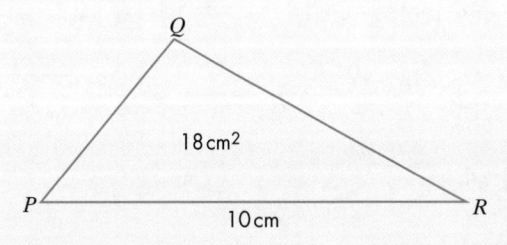
>
> What is the area of triangle XYZ?
>
> The linear scale factor ($PQR \to XYZ$) is $\frac{5}{10} = 0.5$
>
> The area of $XYZ = 18 \times 0.5^2 = 4.5\,\text{cm}^2$

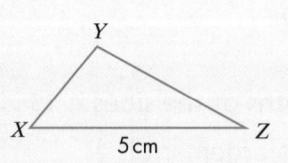

Areas and volumes of similar shapes

You have seen that if the linear scale factor between two similar shapes is k then the area scale factor is k^2.

If two 3D objects have corresponding lengths in the same ratio then their volume ratio is equal to the cube of the linear scale factor. In other words, the **volume scale factor** is k^3.

> **Example**
>
> A manufacturer packages chocolates in two mathematically similar boxes.
>
> The smaller box is 20 cm long and the larger one is 30 cm long. The area of the lid of the smaller one is 250 cm². What is the area of the lid of the larger box?
>
> The linear scale factor (small to big) is $\frac{30}{20} = 1.5$, so the area factor is $1.5^2 = 2.25$
>
> The area of the lid of the larger box is $250 \times 2.25 = 562.5\,\text{cm}^2$
>
> The larger box has a volume of 2531.25 cm³. What is the volume of the smaller box?
>
> The volume ratio is 1.5^3, so the volume of the small box is $2531.25 \div 1.5^3 = 750\,\text{cm}^3$.

> **Revision tip**
>
> The ratio
> length : area : volume
> is expressed as
> $k : k^2 : k^3$

Quick test

1. Use a ruler and a protractor to draw this triangle accurately and then measure the lengths of sides AC and BC.

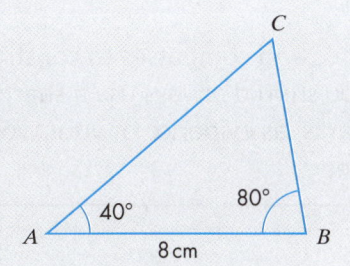

2. An aeroplane travelling from London directly to Paris flies on a bearing of 165°. On what bearing must the pilot fly to return directly to London?

3. Will this net fold to make a cube?

4. Draw a rhombus $ABCD$. Draw the diagonals, labelling the point of their intersection X. How many congruent triangles are there?

5. Alice says that these shapes are mathematically similar.
   Hannah says they are not similar.
   Who is right? Explain your answer.

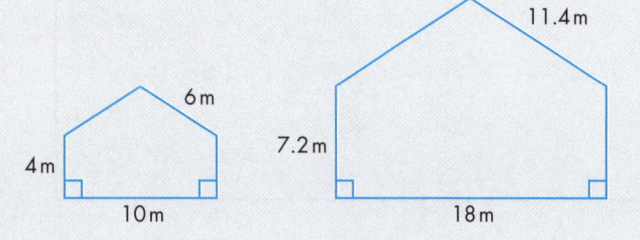

Extended

6. These two triangles are similar.
   What is the area of the smaller triangle?

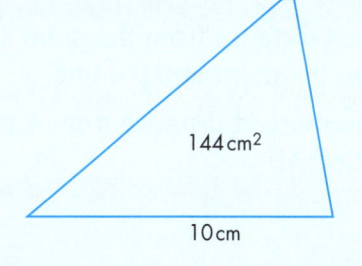

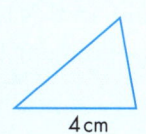

7. Two bottles of fruit juice are mathematically similar. The capacity of the smaller bottle is 500 ml and the larger bottle holds 1.5 litres. The smaller bottle has a base diameter of 6 cm. What is the radius, in mm, of the larger (to 2 s.f.)?

8. Are these two triangles on the right congruent? Explain your answer.

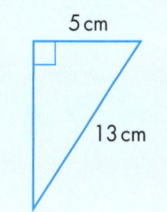

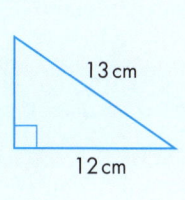

Geometry   97

Geometrical constructions

Constructing a triangle

You should be able to use a ruler, protractor and compasses to construct triangles. Accuracy is very important so you should always use a sharp pencil. Always leave in all construction marks, as evidence of your working, and make sure they are thin, clean and clear.

Example

Draw a triangle with side lengths 5 cm, 4 cm and 6 cm.

Step 1: Take the longest side as being the base. With a ruler draw a line of 6 cm.

Step 2: The next longest side is 5 cm. Open your compasses to 5 cm and placing the compass point at one end of the 6 cm line draw an arc above the line.

Step 3: Open your compasses to the length of the third side, 4 cm, and placing the compass point on the other end of the 6 cm line draw an arc above the line to cross the first arc.

Step 4: Connect each end of the base line to the point where the two arcs intersect.

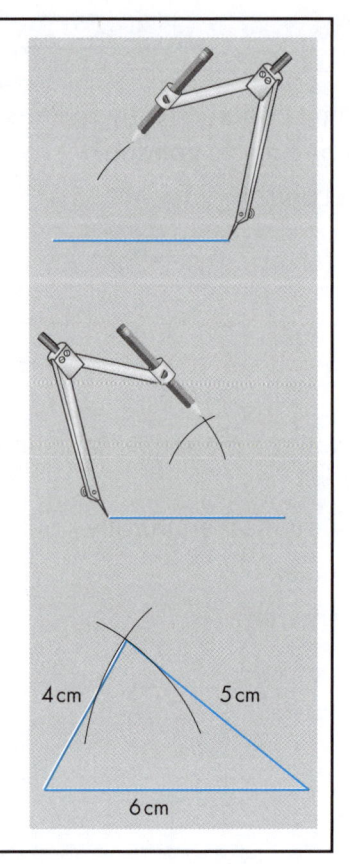

Perpendicular distance

If a point is above a line, shortest distance from the point to the line is the perpendicular distance between the point and the line.

For example, in this triangle the shortest distance from A to the line BC is a line perpendicular to BC making AD.

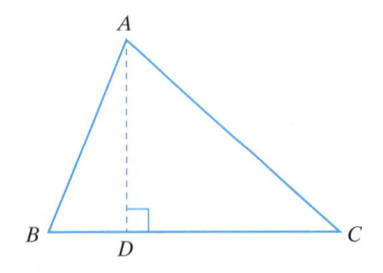

Scale drawings

A **scale drawing** is an accurate representation of a real object, such as an architect's plans for a building. Generally, the drawing is smaller than the object it represents. Sometimes scale drawings are larger, for example, when they represent a microscopic insect or a blood cell.

The scale that is used to make the drawing will be stated, for example, as '1 cm represents 10 m'.

This is a sketch showing the positions of three capital cities. Make an accurate scale drawing, based on the information in the sketch. Use a scale of 1 cm to 150 km. Use the diagram to find the actual distance between London and Madrid.

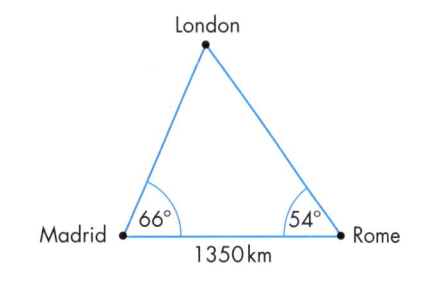

Draw the base line of the triangle:
1350 ÷ 150 = 9 cm

Use a protractor to measure the angles of 66° and 54°, then draw the sides of the triangle and complete it to find the position of London.

Measure the line from London to Madrid. It should be 8.5 cm. Using the diagram the actual distance from London to Madrid is 8.5 × 150 = 1275 km.

Quick test

1. Make an accurate drawing of the kite shown in the sketch on the right.
2. Is it possible to draw a triangle with side lengths: 8 cm, 15 cm and 6 cm? Explain your answer.
3. The sketch shows the angles and approximate distances between four major USA cities. Using a scale of 1 cm = 200 km make an accurate drawing of this map.
   How many kilometres is the distance between Dallas and Chicago?

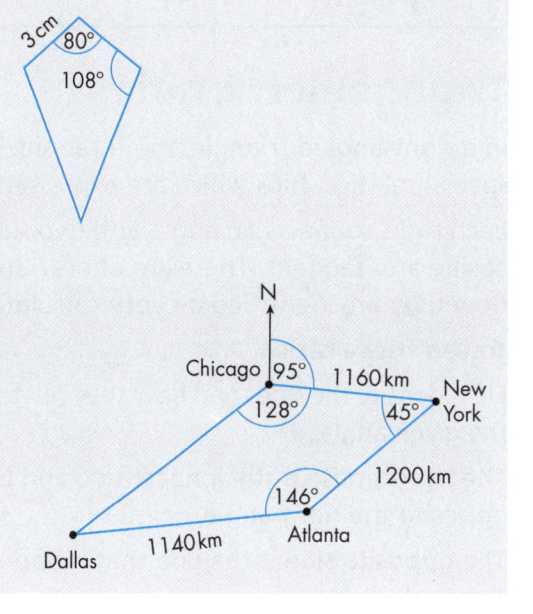

Geometry

Trigonometry

Pythagoras' theorem

The longest side of a right-angled triangle is called the **hypotenuse**. It is always opposite the right-angle.

Pythagoras' theorem states that: in any right-angled triangle, the area of a square drawn on the hypotenuse is equal to the sum of the areas of squares drawn on the other two sides.

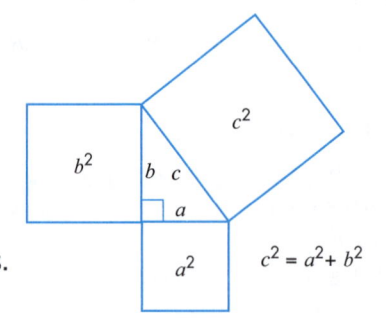

As a formula this is written as $c^2 = a^2 + b^2$

Example

a) Find the length of x.

$x^2 = 5^2 + 8^2$

$x^2 = 89$

$x = \sqrt{89}$

$ = 9.43\,\text{cm}$

(2 d.p.)

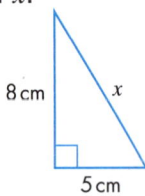

b) Find the length of x.

$17^2 = 12^2 + x^2$

$289 - 144 = x^2$

$145 = x^2$

$x = \sqrt{145} = 12.04\,\text{cm}$ (2 d.p.)

Trigonometric ratios

In a right-angled triangle, the three sides and the three angles share some special relationships which are expressed as ratios.

Each ratio connects an angle and two sides. The three ratios are **sine**, **cosine** and **tangent**. There are abbreviated to sin, cos and tan, which is how they are identified on your calculator.

To use these ratios

First identify the sides and how they relate to the given angle, θ.

The hypotenuse is the longest side and is opposite the right angle.

The **opposite side** is the one that is opposite the angle θ.

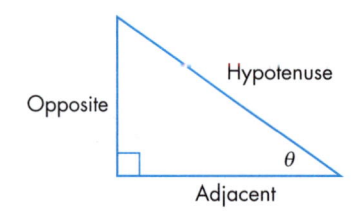

The **adjacent side** has the right angle at one end and θ at the other.

The ratios are defined as:

$\sin \theta = \frac{\text{opposite}}{\text{hypotenuse}} \qquad \cos \theta = \frac{\text{adjacent}}{\text{hypotenuse}} \qquad \tan \theta = \frac{\text{opposite}}{\text{adjacent}}$

These are usually abbreviated to: $\sin \theta = \frac{O}{H} \qquad \cos \theta = \frac{A}{H} \qquad \tan \theta = \frac{O}{A}$

The sin, cos or tan of an angle is a number that is specific to the angle and that can be used in calculations.

To find the sine of 40°, or sin (40°): press [sin] 40 = to get an answer of 0.64278…

Use your calculator to find (to 4 d.p.): sin 35°, cos 56°, tan 72°, 3 sin 65°, 4 cos 31° and 6 tan 10°

You should get: 0.5736, 0.5592, 3.0777, 2.7189, 3.4287 and 1.0580

You need to be able to recall and use sin, cos and tan of 30°, 45° and 60°. Use these diagrams to help you remember them.

	0°	30°	45°	60°	90°
sin	0	$\frac{1}{2}$	$\frac{\sqrt{2}}{2}\left(\frac{1}{\sqrt{2}}\right)$	$\frac{\sqrt{3}}{2}$	1
cos	1	$\frac{\sqrt{3}}{2}$	$\frac{\sqrt{2}}{2}\left(\frac{1}{\sqrt{2}}\right)$	$\frac{1}{2}$	0
tan	0	$\frac{\sqrt{3}}{3}\left(\frac{1}{\sqrt{3}}\right)$	1	$\sqrt{3}$	not defined

Calculating angles

When you know the value of the ratio (sin, cos or tan) of an angle and you need to find the angle itself, you will need to perform the inverse (or opposite) operation.

Suppose, at the end of a calculation you know the cosine of an angle is 0.3456, then the angle is written as $\cos^{-1}(0.3456)$, and you key this into your calculator as 0.3456 [shift] [cos] =, to get an answer of 69.8° (to 1 d.p.).

What angle has a sine of 0.6543 (to 1 d.p.)? $\sin^{-1}(0.6543) = 40.9°$

What angle has a tangent of 1.0789 (to 1 d.p.)? $\tan^{-1}(1.0789) = 47.2°$

Note that $\sin^{-1}(x) + \cos^{-1}(x) = 90°$ and this is true for any value of x.

> **Revision tip**
>
> For Cambridge IGCSE examinations, you need to give angle solutions correct to one decimal place, unless otherwise stated in the question.

Using sine, cosine and tangent functions

The **sine ratio** is $\sin\theta = \frac{\text{opposite}}{\text{hypotenuse}}$

Example

a) In this triangle, the side of length 14 cm is opposite the angle θ and the hypotenuse is 20 cm. Find θ.

To find the angle with sine 0.7, use the inverse sine function:

$\sin^{-1}(0.7) = 44.4°$ (1 d.p.)

b) In this triangle, the side length a is opposite the angle of 32° and the hypotenuse is 6 cm. Find a.

$\sin 32° = \frac{a}{6} \rightarrow a = 6 \times \sin 32°$

$a = 3.18$ cm (2 d.p.)

c) In this triangle, the side of length 8 cm is opposite the angle of 55° and the hypotenuse is b. Find b.

$\sin 55° = \frac{8}{b} \rightarrow b = \frac{8}{\sin 55°} \rightarrow b = 9.77$ cm (2 d.p.)

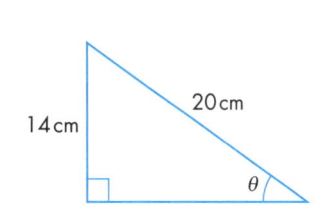

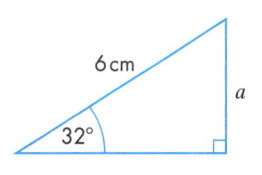

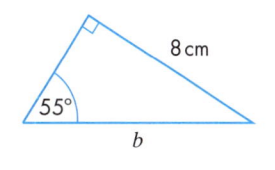

> **Revision tip**
>
> The sine or cosine of any angle is always in the range −1 to +1.

Geometry

The cosine ratio is $\cos \theta = \frac{\text{adjacent}}{\text{hypotenuse}}$

Example

a) In this triangle, the side of length 10 cm is adjacent to the angle θ and the hypotenuse is 15 cm. Find θ.

$\cos \theta = \frac{10}{15} = 0.6667$

To find the angle with cosine 0.6667, use the inverse cosine function.

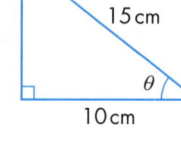

$\cos^{-1}(0.6667) = 48.2°$ (1 d.p.)

b) In this triangle, the side of length c is adjacent to the angle of 42° and the hypotenuse is 12 cm. Find c.

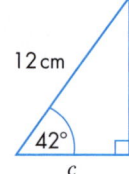

$\cos 42° = \frac{c}{12} \rightarrow c = 12 \times \cos 42° \rightarrow c = 8.92$ cm (2 d.p.)

c) In this triangle, the side of length 9 cm is adjacent to the angle of 28° and the hypotenuse is d. Find d.

$\cos 28° = \frac{9}{d} \rightarrow d = \frac{9}{\cos 28°} \rightarrow d = 10.19$ cm (2 d.p.)

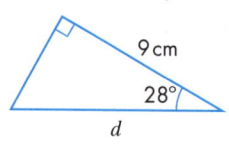

The tangent ratio is $\tan \theta = \frac{\text{opposite}}{\text{adjacent}}$

Example

a) In this triangle, the side length 10 cm is adjacent to the angle θ and the opposite side is 12 cm. Find θ.

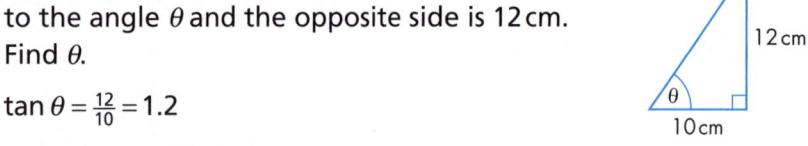

$\tan \theta = \frac{12}{10} = 1.2$

To find the angle with tangent is 1.2, use the inverse tangent function.

$\theta = \tan^{-1} 1.2 = 50.2°$ (1 d.p.)

b) In this triangle, the side of length 15 cm is adjacent to the angle of 34° and and the opposite side is e. Find e.

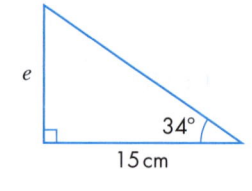

$\tan 34° = \frac{e}{15} \rightarrow e = 15 \times \tan 34° \rightarrow$
$e = 10.12$ cm (2 d.p.)

c) In this triangle, the side of length 9 cm is opposite the angle of 72° and the adjacent side is f. Find f.

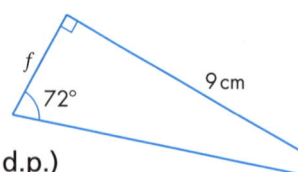

$\tan 72° = \frac{9}{f} \rightarrow f = \frac{9}{\tan 72°} \rightarrow f = 2.92$ cm (2 d.p.)

Which ratio to use

You need to be able to decide which of the three ratios to use.

Try to memorise the acronym SOHCAHTOA. This states the rules, if you say to yourself: 'S(ine) O(pposite divided by) H(ypotenuse),' and so on.

Once you have identified the sides, SOHCAHTOA will help you work out the ratio to use.

> **Revision tip**
>
> Use the acronym SOHCAHTOA to help you remember the three trigonometric ratios and the sides that refer to each ratio.

Example

a) In this triangle, the side of length 13 cm is opposite the angle θ and the side of length 20 cm is the hypotenuse.

The sine ratio uses opposite and hypotenuse: $\sin\theta = \frac{O}{H}$

$\sin\theta = \frac{13}{20} = 0.65$ so $\theta = \sin^{-1}(0.65) = 40.5°$ (1 d.p.)

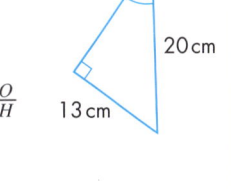

b) In this triangle, the side h is opposite the angle of 18° and side of length 17 cm is adjacent to it.

The tangent ratio uses opposite and adjacent: $\tan\theta = \frac{O}{A}$

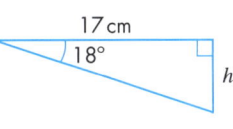

$\tan 18° = \frac{h}{17} \rightarrow h = 17 \times \tan 18° = 5.52$ cm (2 d.p.)

> **Revision tip**
>
> Always show evidence of working – this includes marking on a diagram the sizes of the sides and angles as you identify them.

Extended

Angles of elevation and depression

When you look up at an aeroplane in the sky, or look down from the top of a sea cliff at someone on the beach, your eye line is moving away from the horizontal. In each case your new eye line makes an angle with the horizontal.

When you look up at something you create an **angle of elevation**.

When you look down, you create an **angle of depression**.

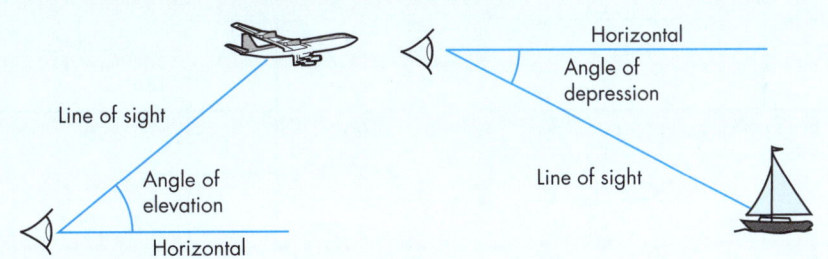

Sami is lying on the ground, looking up at a flag that is hanging at the top of a 100 m mast. He is 40 m away from the foot of the mast. What is the angle of elevation from Sami to the top of the flag?

The side 100 m is opposite the angle and the side 40 m is adjacent to it.

So $\tan\theta = \frac{100}{40}$ and $\theta = \tan^{-1}(2.5) \approx 68.2°$ (1 d.p.)

Geometry

Problems in three dimensions

This cuboid has dimensions of 4 cm, 5 cm and 14 cm.

Find the lengths EG, AG and Angle AGE.

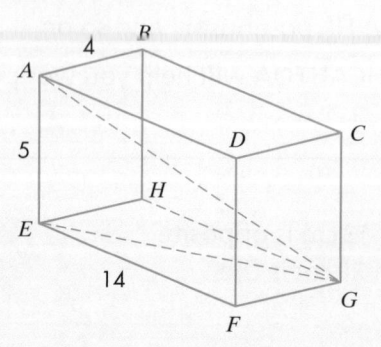

> **Revision tip**
>
> When working with 3D questions, find a right-angled triangle in the figure that includes two known values which you can use in a calculation.

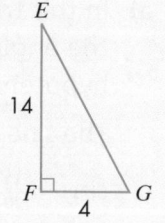

Draw small sketches of the triangles within the cuboid, that include lengths or angles that you need to use.

$EG = \sqrt{EF^2 + FG^2} = \sqrt{14^2 + 4^2} = \sqrt{212} = 14.56$ cm (2 d.p.)

$AG = \sqrt{AE^2 + EG^2} = \sqrt{5^2 + 212} = \sqrt{237} = 15.39$ cm (2 d.p.)

$\tan \theta = \dfrac{5}{14.46} \rightarrow \theta = \tan^{-1}(0.3434) = 19.0°$ (1 d.p.)

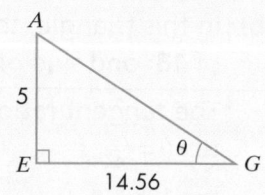

Sine and cosine of obtuse angles

So far, you have worked with the sines and cosines of angles in right-angled triangles, which means that the angles have been acute.

Obtuse angles also have sine and cosine values.

On your calculator, find sin 135° and cos 135°, to 4 decimal places: you should get 0.7071 and −0.7071

These are the sine and cosine curves, or waves, the graphs produced by plotting the sine or cosine of an angle.

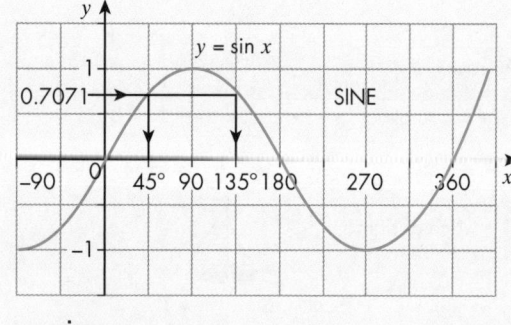

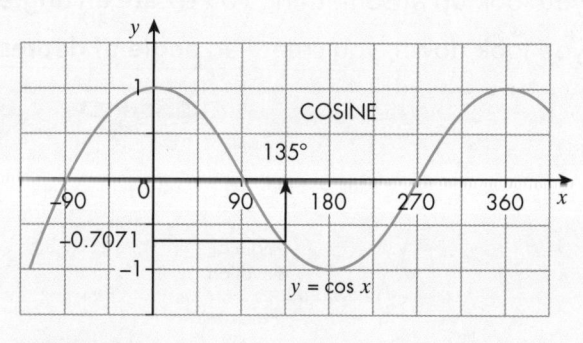

$y = \sin x$

$y = \cos x$

The largest and smallest values that sine and cosine can take are +1 and −1

For sine, all angles in the range $0 \leq x \leq 180°$ have a positive value and all angles in the range $180 < x < 360°$ have a negative value.

From the graph, you can see that sin 45° = sin 135° = 0.7071, meaning $\sin \theta = \sin (180° - \theta)$

For cosine, all angles in the range $0 \leq x \leq 90°$ have a positive value and all angles in the range $90° < x < 270°$ have a negative value.

Cosines of θ and $(180° - \theta)$ have the same numeric value but the cosine of the obtuse angle is negative: cos 135° = −cos 45° = −0.7071

This is important to know if a question asks you to identify the angles with a particular sine or cosine value.

For example, if $\cos x = \frac{\sqrt{2}}{2}$, then x is returned as 45°. However, the shape of the curve tells you that x could also be 315° (360° − 45°). Check, $\cos(315) = \frac{\sqrt{2}}{2}$

Additionally, a question might be looking for an answer that is obtuse but the calculator returns an acute angle. Understanding the shape of the curves helps you see how to obtain the obtuse angle given a particular sine or cosine value.

You can see from the two graphs that a particular sine or cosine value can return reflex angles as well as the obtuse angle.

For example, a sine of −0.342 returns an angle of 200°, or 340°. A cosine of −0.7071 also returns an angle of 225° and + 0.7071 also returns an angle of 315°.

The graph of tangent has a very different shape to the graphs of sine and cosine.

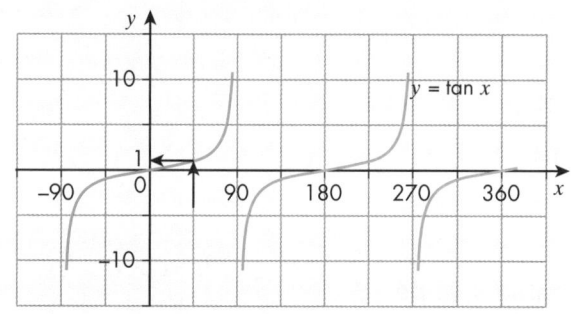

It has asymptotes at −90° and 90° and the pattern repeats every 180°. $\tan(45°) = 1$, but $\tan(x°) = 1$ could be 45°, or 45 + 180 = 225°

The sine rule and the cosine rule

When you need to work out the sizes of angles or sides in a triangle that does not have a right-angle, there are two further rules that you can use: the **sine rule** and the **cosine rule**.

In any triangle, the connection of an angle to its opposite side is important. Remember the standard way of labelling the sides and angles.

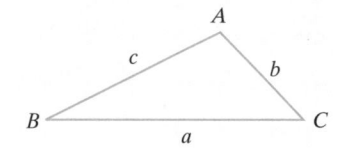

Then the sine rule is: $\frac{a}{\sin A} = \frac{b}{\sin B} = \frac{c}{\sin C}$

When you need to find an angle, invert this rule so that the sines are on top.

$\frac{\sin A}{a} = \frac{\sin B}{b} = \frac{\sin C}{c}$

> **Revision tip**
>
> Any triangle has six measurements: three sides and three angles. To find any unknown angle or side, you need to know at least three other pieces of information.

To work out the length of a in this triangle:

$\frac{a}{\sin 84°} = \frac{4}{\sin 58°} \rightarrow a = \frac{4 \sin 84°}{\sin 58°} = 4.69$ cm (2 d.p.)

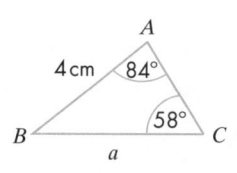

To work out the angle θ in this triangle:

$\frac{\sin 28°}{6.5} = \frac{\sin \theta}{12} \rightarrow \sin \theta = \frac{12 \sin 28°}{6.5} = 0.8776$

$\sin^{-1}(0.8667) = 60.1°$ (1 d.p.)

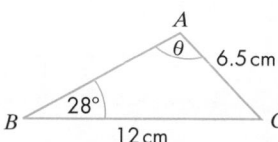

Geometry

Work out the angle θ in this triangle:

$$\frac{\sin 15°}{6} = \frac{\sin \theta}{20} \rightarrow \sin \theta = \frac{20 \sin 15°}{6} = 0.8623$$

$\sin^{-1}(0.8623) = 59.6°$ (1 d.p.)

However, $\theta > 90°$

Remember that $\sin^{-1}(0.8623)$ can lead to more than one angle, so to find the obtuse angle, $\sin^{-1}(0.8623) = 180° - 59.6° = 120.4°$ (1 d.p.)

The cosine rule is: $a^2 = b^2 + c^2 - 2bc \cos A$

Work out the length of x in this triangle:

$x^2 = 9^2 + 10^2 - 2 \times 9 \times 10 \times \cos 34°$

$x^2 = 31.7732$

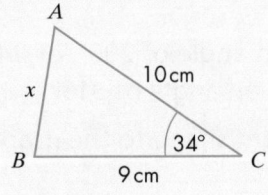

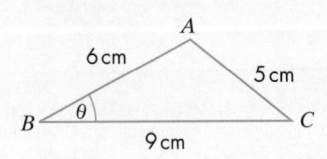

$x = 5.64$ cm (2 d.p.)

In this triangle on the left:

$5^2 = 9^2 + 6^2 - 2 \times 9 \times 6 \times \cos \theta$

Rearrange $\rightarrow \cos \theta = \frac{9^2 + 6^2 - 5^2}{2 \times 9 \times 6} = 0.8518$

$\theta = \cos^{-1}(0.8518) = 31.6°$ (1 d.p.)

Using sine to find the area of a triangle

The rule, area = $\frac{1}{2}$ base × height, can be adjusted slightly to enable it to be used when you don't know the **vertical height**, h. The **area rule** is:

area = $\frac{1}{2} ab \sin C$

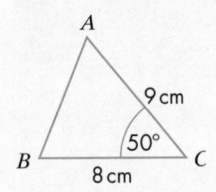

In this triangle:

area of $ABC = \frac{1}{2} \times 8 \times 9 \times \sin 50° = 27.58$ cm² (2 d.p.)

To work out length c in this triangle on the right:

$18 = \frac{1}{2} \times 7 \times c \times \sin 40° \rightarrow c = \frac{18}{\frac{1}{2} \times 7 \times \sin 40°} = 8.00$ cm (2 d.p.)

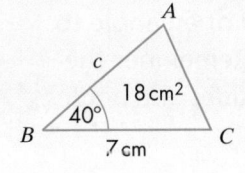

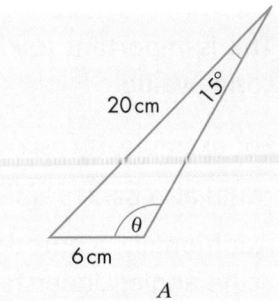

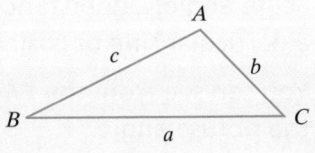

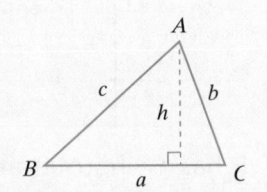

Quick test

1. In a right-angled triangle, the lengths of two sides are 6 cm and 9 cm. How long is the hypotenuse? Give your answer in cm and to 2 d.p.
2. A right-angled triangle has a side of length 4 cm and its hypotenuse is 15 cm. What is the length of the third side? Give your answer in cm and to 2 d.p.
3. Find sin 56°, cos 56° and tan 56°
4. Work out the angle θ (to the nearest degree) when **a)** tan θ = 2.475 **b)** sin θ = 0.5592
5. Work out the missing side lengths x and y (to 2 d.p.) and the angle θ (to 1 d.p.).

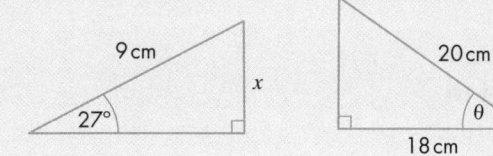

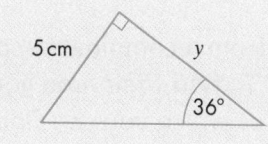

Extended

6. A pilot, crossing the Atlantic Ocean, looks out of the cockpit and sees the coast of the USA, which is 15 km away. The angle of depression from the pilot to the coast is 18°. At what height, in metres, is the aeroplane (to 4 s.f.)?
7. $XABCD$ is a square-based pyramid. Calculate the size of Angle XDB (to 1 d.p.).

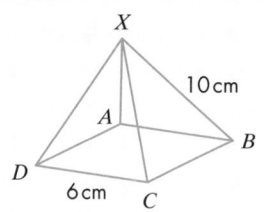

8. Find the size of these obtuse angles. Give your answers to 1 d.p.

    a) $\cos^{-1}(-0.2588)$ **b)** $\sin^{-1}(0.1736)$

9. Work out the angle θ (to 1 d.p.) and the area of the triangle (to 2 d.p.).

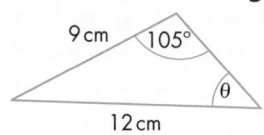

10. Find the length of the third side of this triangle.

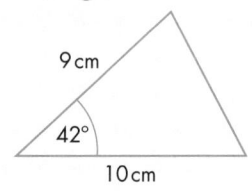

Mensuration

Perimeter and area of a rectangle

The **perimeter** of any two-dimensional shape is the total distance around the edge. The **area** is the amount of space inside the perimeter.

In a rectangle:

perimeter $= l + w + l + w = 2(l + w)$

area $= l \times w$

Shapes that are made from a combination of shapes are compound shapes. To find their area you need to split them into shapes of which you can find the areas and then add these separate areas together.

For example, consider shape X on the right.

First, find the missing dimensions and then split the shape into smaller rectangles.

perimeter $= 4 + 4 + 4 + 5 + 1 + 1 + 9 + 10 = 38$ cm

area $A = 4 \times 10 = 40$

area $B = 4 \times 6 = 24$

area $C = 1 \times 1 = 1$

total area $= 40 + 24 + 1 = 65$ cm^2

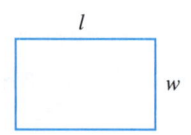

Revision tip

Make sure you know the difference between perimeter and area and are confident identifying the measurements you need to work them out.

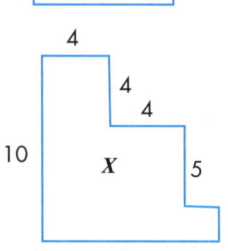

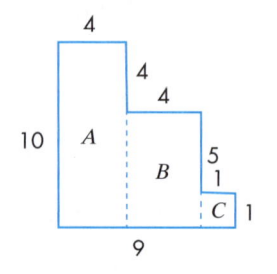

Area of a triangle

The area of any triangle is:
area $= \frac{1}{2} \times$ base $\times$ perpendicular height

Or more usually: $A = \frac{1}{2}bh$

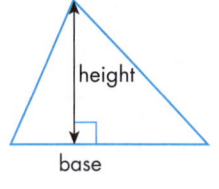

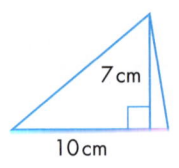

area $= \frac{1}{2} \times 10 \times 7$
$= 35$ cm^2

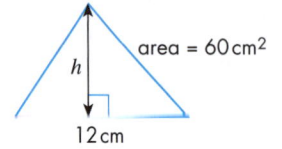

$60 = \frac{1}{2} \times 12 \times h$

$\frac{60}{6} = h$

$h = 10$ cm

Area of a parallelogram

The area of a parallelogram $=$ base $\times$ height

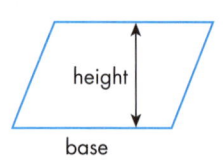

area $= 2 \times 9 = 18$ cm^2

parallelogram area $= 10 \times 7 = 70$ cm^2

triangle area $= \frac{1}{2} \times 6 \times 4 = 12$ cm^2

area of shaded region $= 70 - 12 = 58$ cm^2

Area of a trapezium

The area of trapezium is $A = \frac{1}{2}(a+b) \times h$ where a and b are the lengths of the parallel sides and h is the perpendicular distance between them.

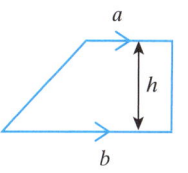

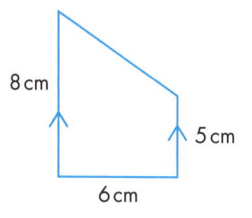

area = $\frac{1}{2}(5+8) \times 6$
area = $39\,\text{cm}^2$

$72 = \frac{1}{2}(4+12) \times h$
$72 = 8h$
$h = 9\,\text{cm}$

Circumference and area of a circle

Read the question carefully to identify how you need to present the answer. You will need to round calculations with π, using d.p. or s.f., unless you are told to leave them in terms of π

The perimeter of a circle is called the **circumference**.

The circumference, C, of a circle, with diameter d is $C = \pi d$

Since the diameter is twice the radius, r, the circumference can also be given as $C = 2\pi r$

The area, A, of a circle with, radius r, is $A = \pi r^2$

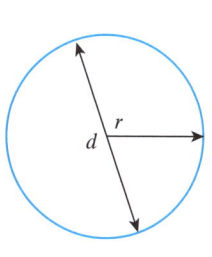

On your calculator look for [π ×10^x].

In this case you would need to press [shift] first.

If you do not have π on your calculator use the value 3.142

You will be given this value for π on the front cover of your exam booklet.

Example

a) Find the area and circumference of this circle (to 2 d.p.).

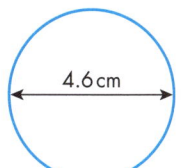

circumference = πd
$C = \pi \times 4.6$
$C = 14.45\,\text{cm}$

b) Find the area of the shaded region, leaving your answer in terms of π

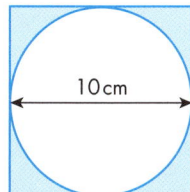

area of the square = $10^2 = 100$
area of the circle = $\pi \times 5^2 = 25\pi$
area of the shaded region = $100 - 25\pi\,\text{cm}^2$

> **Revision tip**
>
> Remember you should always give the units with your answer to a question.

Arcs and sectors

A **sector** is a fraction of a circle, bounded by two radii and an arc.

An **arc** is a fraction of the circumference.

The fraction of the area or circumference to use is determined by the size of the angle in the sector, θ. The fraction of the circle which is shaded in the diagram is $\frac{\theta}{360}$

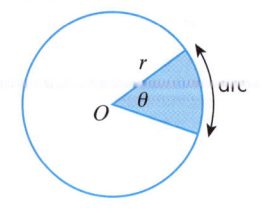

The length of an arc is $\frac{\theta}{360} \times 2\pi r$ or $\frac{\theta}{360} \times \pi d$

The area of a sector is $\frac{\theta}{360} \times \pi r^2$

For the shaded sector in circle X:

arc length $= \frac{120}{360} \times 2 \times \pi \times 6 = 4\pi$ or $12.57\,\text{cm}$ (2 d.p.)

sector area $= \frac{120}{360} \times \pi \times 6^2 = 12\pi$ or $37.7\,\text{cm}^2$ (3 s.f.)

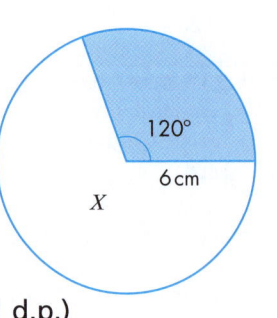

perimeter $= \left(\frac{120}{360} \times 2 \times \pi \times 6\right) + 6 + 6 = 4\pi + 12$ or $24.6\,\text{cm}$ (1 d.p.)

> **Revision tip**
>
> You may be asked to leave answers 'in terms of π' instead of rounding to d.p. or s.f., so make sure you read the question carefully.

Surface area and volume of a cuboid

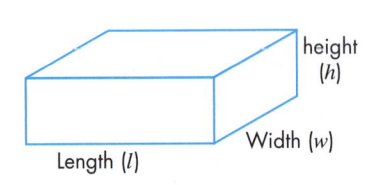

The volume of a cuboid is the product of length, width and height.

volume = length × width × height, $V = lwh$

The **surface area** is the sum of the areas of all six faces.

The opposite rectangular faces in each pair have the same area.

total area of the top and bottom rectangles $= 2 \times l \times w = 2lw$

total area of the front and back rectangles $= 2 \times l \times h = 2lh$

total area of the two side rectangles $= 2 \times w \times h = 2wh$

total surface area $= 2lw + 2lh + 2wh$

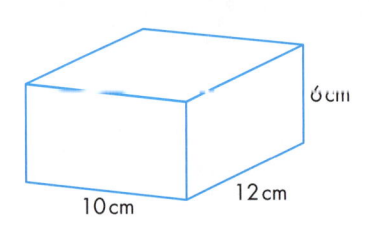

For this cuboid:

volume $= 10 \times 12 \times 6 = 720\,\text{cm}^3$

surface area $= (2 \times 10 \times 12) + (2 \times 10 \times 6) + (2 \times 12 \times 6)$

$= 240 + 120 + 144$

$= 504\,\text{cm}^2$

> **Revision tip**
>
> When finding surface area, remember that you need to find the areas of all the surfaces, not just those you can see.

Volume of a prism

A prism is any solid object that has the same cross-sectional shape throughout its length.

Name	cuboid	triangular prism	cylinder	cuboid	hexagonal prism
3D					
Cross-section	rectangle	triangle	circle	square	hexagon

110 IGCSE Mathematics Revision Guide

To find the volume of a prism, multiply the area of the cross-section by the length of the object.

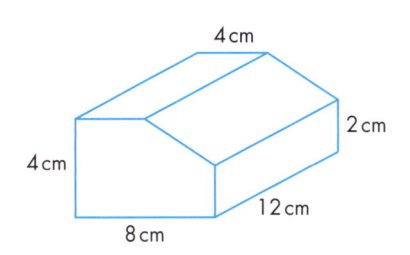

volume = area of cross-section × length

For this prism:

first calculate the area of the cross-section by splitting it into separate shapes.

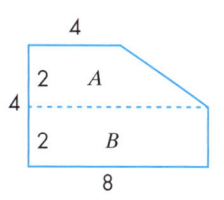

area of $A = \frac{1}{2}(4 + 8) \times 2 = 12$

area of $B = 8 \times 2 = 16$

cross-sectional area = 12 + 16 = 28 cm²

volume = 28 × 12 = 336 cm³

Volume and surface area of a cylinder

A cylinder is a prism in which the cross-section is a circle.

The same formula applies.

volume = area of cross-section × length

volume = $\pi r^2 l$

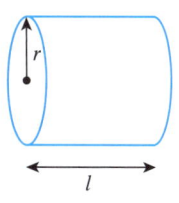

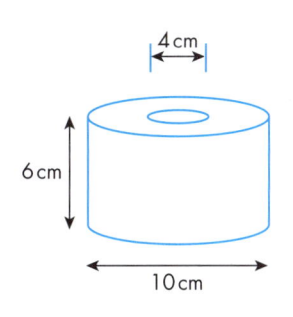

This cylinder has an outer diameter of 10 cm. A circular hole with diameter of 4 cm has been drilled through the middle, along its length. What is the volume of this object (to 3 s.f.)?

area of the whole circular end = $\pi \times 5^2 = 25\pi$

area of the circular hole = $\pi \times 2^2 = 4\pi$

area of the solid cross-section = $25\pi - 4\pi = 21\pi$

volume of the object = $21\pi \times 6 = 126\pi = 396$ cm³ (3 s.f.)

> **Revision tip**
>
> Read the question carefully. When working with cylinders and cones, you may be asked only to find the curved surface area.

The total surface area of a cylinder is the area of the two circular ends plus the curved surface area.

The curved surface area is a rectangle with length equal to the circumference of the end circle.

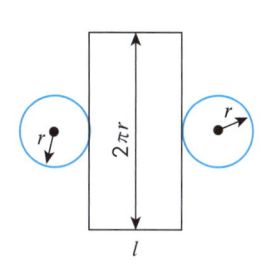

Total surface area = $2\pi r^2 + 2\pi r l$

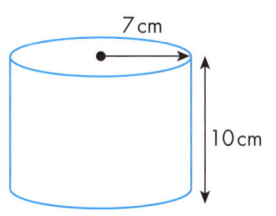

What is the total surface area of this cylinder, to 3 s.f.?

area of the two circular ends = $2 \times \pi \times 7^2 = 98\pi$

curved surface area = $2 \times \pi \times 7 \times 10 = 140\pi$

total surface area = $238\pi = 748$ cm² (3 s.f.)

Geometry

Volume of a pyramid

A pyramid has a base, which may be a polygon with any number of sides, with triangular faces meeting at a common vertex.

The volume of a pyramid is: volume = $\frac{1}{3}$ × base area × vertical height

Example: This square-based pyramid has volume 60 cm³ and vertical height 5 cm.

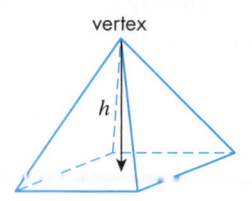

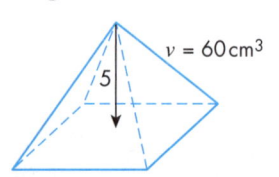

What is the side length of the square base?

$60 = \frac{1}{3}$ × base area × 5

base area = 36 cm²

side length = $\sqrt{36}$ = 6 cm

Volume and surface area of a cone

A cone is a special kind of pyramid. It has a circular base, so the formula given above for prisms can be applied to a cone.

Volume = $\frac{1}{3}\pi r^2 h$ where r is the radius of the base, and h is the perpendicular height of the cone.

It is often helpful to use the slant height of a cone. This is shown as l on the diagram. The area of the curved surface of the cone would open up into a sector of a larger circle with radius l.

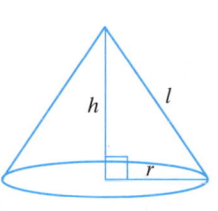

Curved surface area = $\pi r l$

Total surface area of the cone = the circular base + the curved surface area
= $\pi r^2 + \pi r l$

> **Revision tip**
>
> Questions involving pyramids and cones may introduce a shape called a **frustum**. Make sure you know what this looks like and how to calculate its volume.

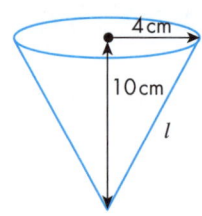

An ice-cream cone has radius 4 cm and depth 10 cm.

How much ice-cream will it hold? Give your answer to the nearest ml.

What is the area of the curved surface of the cone? Give your answer to 2 d.p.

volume = $\frac{1}{3} \times \pi \times 4^2 \times 10$ = 167.6 = 168 cm³ = 168 ml (to nearest ml)

To find the area of the curved surface, you need to use Pythagoras' theorem to find the slant height, l.

$l = \sqrt{10^4 + 4^2} = \sqrt{116}$ = 10.77

area of curved surface = $\pi \times 4 \times 10.77$ = 135.34 cm² (2 d.p.)

Volume and surface area of a sphere

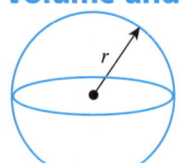

Volume = $\frac{4}{3}\pi r^3$

Surface area = $4\pi r^2$

A glass **sphere** has a diameter of 22 cm. It fits exactly into a cubical presentation box.
What is the surface area of the sphere? Give your answer to 3 s.f.
surface area = $4 \times \pi \times 11^2$ = 1520 cm²
The sphere is placed in the box and then the box is filled with foam granules to protect it.
What volume of granules, to the nearest ml, is required to fill the empty space around the sphere inside the box?
volume of box = 22^3 = 10648 cm³
volume of sphere = $\frac{4}{3} \times \pi\, 11^3$ = 5575.28 cm³
volume of granules = 10648 − 5575.28 = 5073 cm³ = 5073 ml (to nearest ml)

IGCSE Mathematics Revision Guide

Quick test

1. Find the area of this shape. (All dimensions are in cm.)

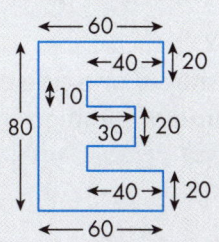

2. Find the area of this net for a square-base pyramid, to 2 s.f.

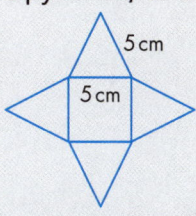

3. This parallelogram has a smaller parallelogram cut from it. What is the area of the shaded region?

4. This trapezium has an area of 45 cm². What is its height, h?

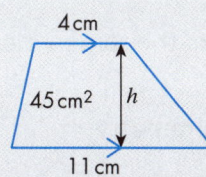

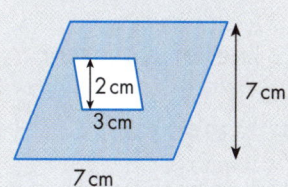

5. A circle has a circumference of 28.27 cm. What is its area (to 3 s.f.)?

6. A cuboid box holds 4200 cm³ of breakfast cereal. It has a rectangular base measuring 20 cm × 6 cm. How tall is the box?

7. Some mathematical signs are being made from wood. They have the dimensions as in this diagram.
   If the wood that the sign is being made from has a density of 0.7 g/cm³, what is the mass of the sign?

8. A wax solid cylinder has a diameter of 6 cm and a length of 20 cm. It is melted down and recast into a cube. What is the side length of the cube? Give your answer to 2 d.p.

9. Calculate the total perimeter and area of this sector. Give your answers in terms of pi.

10. The shaded part of this sector is called a segment. What is its area, to 2 d.p.?

11. A metal pyramid has a rectangular base, 3 cm × 4 cm, and its height is 18 cm.
    It is melted down and recast as a cuboid with the same base area as the pyramid. What is its height?

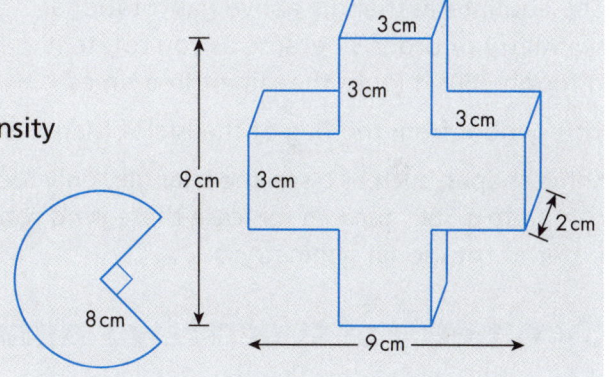

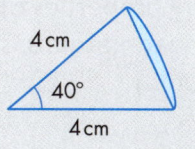

Extended

12. A frustum of a cone has the dimensions, in cm, as shown.
    What is its volume, to 3 s.f.?

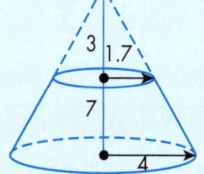

13. A hemisphere is made out of plaster. It has a diameter of 200 cm. What is its total surface area, in metres, to 4 s.f.?

Geometry 113

Symmetry

Lines of symmetry

When a line of symmetry is drawn through a shape, one side of the shape is a **reflection** of the other side. This is why a line of symmetry is often called a mirror line.

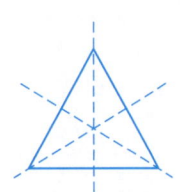

1 line of symmety   3 lines of symmetry

Rotational symmetry

A shape has **rotational symmetry** when it looks the same when **rotated** about a central point.

The best way to find the **order of rotational symmetry** is to use tracing paper.

The order of symmetry is the number of times the shape looks the same as it is turned.

The equilateral triangle above has rotational symmetry of order 3 because as you rotate it through 360° it looks the same three times.

This symbol from the flag of the Isle of Man has rotational symmetry of 3.

Some shapes, such as a **scalene triangle**, only look the same when you complete a 360° turn. In this case there is no rotational symmetry so the order of rotational symmetry is 1.

Symmetry of special two-dimensional shapes

Some three- and four-sided shapes have special names and properties. You should know about the symmetry properties of isosceles and equilateral triangles, as well as the special quadrilaterals: square, rectangle, rhombus, parallelogram, kite and trapezium.

For example:

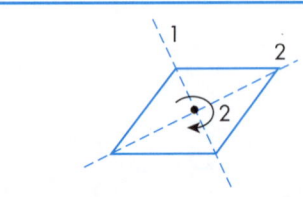

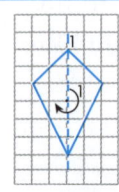

A rhombus has two lines of symmetry and rotational symmetry of order 2.	A kite has one line of symmetry and no rotational symmetry (order 1).

IGCSE Mathematics Revision Guide

Extended

Symmetry of three-dimensional shapes

Three-dimensional objects may have two types of symmetry: planes of symmetry and axes of symmetry.

If you cut an object in half so that one half is a reflection of the other you have created a plane of symmetry. This triangular prism has one plane of symmetry, as shown. If the triangle was isosceles or equilateral, then there would be planes of symmetry along the length of the prism, matching the lines of symmetry of the triangular faces.

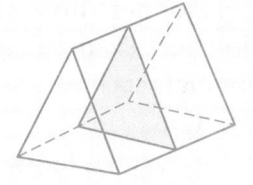

Imagine a piece of wire pushed through an object and spinning the object around the wire. If the shape looks the same in two or more positions, the wire is an axis of symmetry.

This square-based pyramid has an axis of symmetry because, as you turn it around the axis, it will look the same 4 times – it has rotational symmetry of order 4 about the axis.

Just as two-dimensional shapes have lines of symmetry and rotational symmetry, three-dimensional shapes can have planes of symmetry and axes of symmetry.

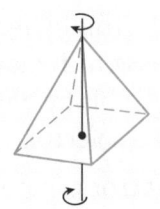

Symmetry in circles

Here are some more facts about circles.

When two chords are the same length, they are the same distance away from the centre of the circle.

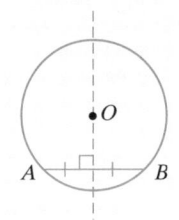

The perpendicular bisector of a chord passes through the centre of the circle.

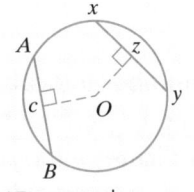

$AB = xy$ and $Oc = Oz$

> **Revision tip**
>
> You can use these new facts, together with those from the chapter on Mensuration, to answer questions involving angles in circles.

Two tangents drawn from a common point are equal in length.

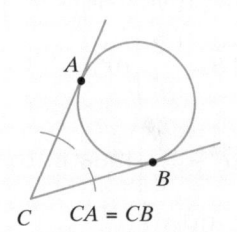

$CA = CB$

Quick test

1. Copy this diagram and then shade in four more squares so that it has 2 lines of symmetry.
2. Write out the alphabet in capital letters. How many letters have rotational symmetry?
3. Gerry says that the only difference between a parallelogram and a rectangle is the fact that a rectangle has right angles and a parallelogram does not. He is wrong, there is another difference. What is it?

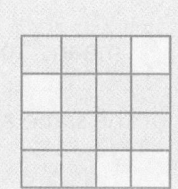

Extended

4. A tetrahedron is a solid object with four faces that are equilateral triangles. How many axes of symmetry does it have?
5. O is the centre of the circle. AB and AC are tangents.
   What is the size of Angle BEC?

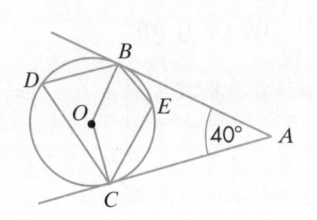

Vectors

Introduction to vectors

A **vector** describes movement: it has **magnitude** and direction. Examples are velocity, acceleration and force. Numbers such as 3 and 17 are scalar, they do not have direction.

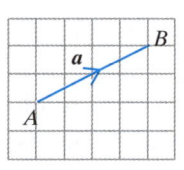

Vectors can be written in different ways.

$\overrightarrow{AB}$ This vector represents a movement from point A to point B, the arrow indicates the direction.

a This lowercase letter, printed in **bold**, represents a vector movement, which may be defined in a diagram. When writing a vector like this by hand, underline it, as in $\underline{a}$, to indicate the letter represents a vector.

A vector can be shown on a coordinate grid. Then it can be represented by two numbers written in a column inside brackets, as $\begin{pmatrix} x \\ y \end{pmatrix}$

> **Revision tip**
>
> Do not forget the arrow to indicate the vector and its direction.

This is a column vector. The top number indicates movement in the x-direction, along or parallel to the x-axis, and the bottom number indicates a movement in the y-direction, along or parallel to the y-axis. Positive values indicate movement right or up, and negative values indicate movement left or down.

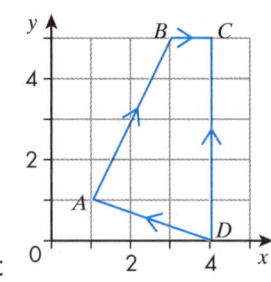

$\overrightarrow{AB} = \begin{pmatrix} 2 \\ 4 \end{pmatrix}$ means 'to get from A to B, move 2 right and 4 upwards'.

$\overrightarrow{BC} = \begin{pmatrix} 1 \\ 0 \end{pmatrix}$ means 'to get from B to C move right 1 without moving up or down'.

$\overrightarrow{DC} = \begin{pmatrix} 0 \\ 5 \end{pmatrix}$ means 'to get from D to C move 5 upwards without moving left or right'.

$\overrightarrow{DA} = \begin{pmatrix} -3 \\ 1 \end{pmatrix}$ means 'to get from D to A move 3 left and 1 upwards'.

$\overrightarrow{BA}$ is not the same as $\overrightarrow{AB}$ because the move from B to A would be $\begin{pmatrix} -2 \\ -4 \end{pmatrix}$

This shows that $\overrightarrow{BA} = -\overrightarrow{AB}$. Similarly, if vector $\mathbf{s} = \begin{pmatrix} -3 \\ 1 \end{pmatrix}$, then $-\mathbf{s} = \begin{pmatrix} 3 \\ -1 \end{pmatrix}$

Vectors can be added, subtracted and multiplied.

In the diagram:

$\mathbf{a} + \mathbf{b} = \mathbf{x}$ or $\begin{pmatrix} 1 \\ 4 \end{pmatrix} + \begin{pmatrix} 3 \\ -1 \end{pmatrix} = \begin{pmatrix} 4 \\ 3 \end{pmatrix}$

$\mathbf{y} - \mathbf{b} = \mathbf{c} \rightarrow \mathbf{y} + (-\mathbf{b}) = \mathbf{c}$ or $\begin{pmatrix} 2 \\ -3 \end{pmatrix} + \begin{pmatrix} -3 \\ 1 \end{pmatrix} = \begin{pmatrix} -1 \\ -2 \end{pmatrix}$

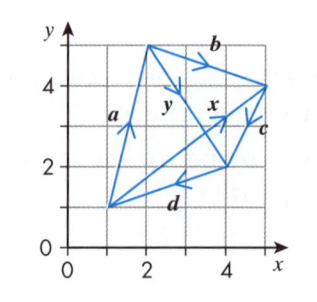

Write **c + d** as a column vector.

$\begin{pmatrix}-1\\-2\end{pmatrix}+\begin{pmatrix}-3\\-1\end{pmatrix}=\begin{pmatrix}-4\\-3\end{pmatrix}$, which is −**x**

Vectors can be multiplied by scalars.

Here, $\mathbf{m}=\begin{pmatrix}-2\\1\end{pmatrix}$ so $2\mathbf{m}=2\cdot\begin{pmatrix}-2\\1\end{pmatrix}=\begin{pmatrix}-4\\2\end{pmatrix}$

In general, if k is a scalar, then $k\begin{pmatrix}x\\y\end{pmatrix}=\begin{pmatrix}kx\\ky\end{pmatrix}$

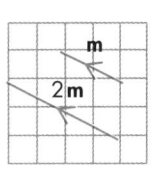

Extended

Using vectors

A position vector is a vector that gives the position of a point, relative to a fixed point (the origin).
In this diagram O is the origin and C is the midpoint of AB.
To work out the vectors $\overrightarrow{AB}$ and $\overrightarrow{AC}$:

$\overrightarrow{AB}=\overrightarrow{AO}+\overrightarrow{OB}=4\mathbf{p}+3\mathbf{q}$

$\overrightarrow{AC}=\tfrac{1}{2}\overrightarrow{AB}=\tfrac{1}{2}(4\mathbf{p}+3\mathbf{q})=2\mathbf{p}+\tfrac{3}{2}\mathbf{q}$

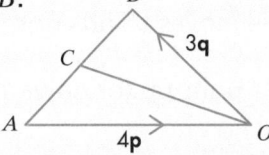

Example

If point D is the midpoint of OB in the triangle above, find the vector $\overrightarrow{CD}$.
Use your findings to describe the relationship between CD and AO.

$\overrightarrow{CD}=\overrightarrow{CB}+\tfrac{1}{2}\overrightarrow{BO}=\overrightarrow{CB}-\tfrac{1}{2}\overrightarrow{OB}$

$\overrightarrow{CB}=\overrightarrow{AC}$

So $\overrightarrow{CD}=2\mathbf{p}+\tfrac{3}{2}\mathbf{q}-\tfrac{1}{2}\times 3\mathbf{q}=2\mathbf{p}+\tfrac{3}{2}\mathbf{q}-\tfrac{3}{2}\mathbf{q}=2\mathbf{p}$

$\overrightarrow{AO}=4\mathbf{p}=2(2\mathbf{p})=2\overrightarrow{CD}$

This means that AO is twice as long as CD and that, because both vectors are multiples of the same vector, **p**, AO and CD are parallel.

> **Revision tip**
>
> Questions will often use a midpoint or divide a line according to a ratio. Take care, when working out the length of the section of the line to use.

The magnitude of a vector

The size or magnitude of a vector is its length.
It is represented by vertical lines either side of the vector, $|\overrightarrow{CD}|$ or a.
When a vector is drawn on a grid you can use Pythagoras' theorem to calculate its magnitude.

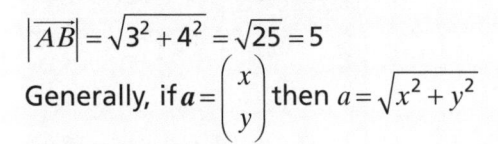

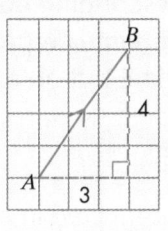

$|\overrightarrow{AB}|=\sqrt{3^2+4^2}=\sqrt{25}=5$

Generally, if $\mathbf{a}=\begin{pmatrix}x\\y\end{pmatrix}$ then $a=\sqrt{x^2+y^2}$

Quick test

1. $\mathbf{m}=\begin{pmatrix}2\\-3\end{pmatrix}$ and $\mathbf{n}=\begin{pmatrix}1\\2\end{pmatrix}$

   Work out these vectors.

   i) $\mathbf{m}+\mathbf{n}$ ii) $3\mathbf{n}$ iii) $\mathbf{m}-\mathbf{n}$ iv) $\mathbf{n}-\mathbf{m}$ v) $2\mathbf{m}+\mathbf{n}$

Extended

2. In the diagram point C divides BD in the ratio 3 : 1. Work out these vectors.

   a) $\overrightarrow{AC}$ b) $\overrightarrow{AE}$ c) $\overrightarrow{BE}$ d) $\overrightarrow{AD}$

3. When the diagram in Q2 is drawn on a coordinate grid, vertex B is at (1, 2), vertex A is at (2, 6) and point D is at (9, 2). Work out $|\overrightarrow{AD}|$ and $|\overrightarrow{BA}|$.

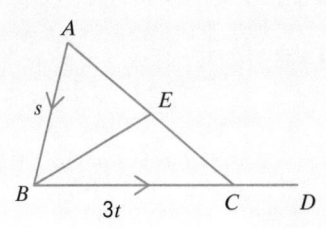

Geometry

Transformations

A **transformation** changes the position or the size of a shape. You need to know about four ways to **transform** a two-dimensional shape: **translation**, reflection, rotation, enlargement. The original shape is the **object** and the transformed shape is the **image**.

Translation

A translation moves a shape without reflecting or rotating it. The image is congruent to the original object; it just slides from one position to another.

The movement or slide can be described with a vector, $\begin{pmatrix} x \\ y \end{pmatrix}$, where the shape moves x units right (or left if x is negative) and y units up (or down if y is negative).

The move from D to B is represented by $\begin{pmatrix} 1 \\ 2 \end{pmatrix}$, from C to D is $\begin{pmatrix} 2 \\ 1 \end{pmatrix}$, from A to B is $\begin{pmatrix} 3 \\ 0 \end{pmatrix}$, from C to A is $\begin{pmatrix} 0 \\ 3 \end{pmatrix}$, from A to D is $\begin{pmatrix} 2 \\ -2 \end{pmatrix}$ and from B to C is $\begin{pmatrix} -3 \\ -3 \end{pmatrix}$.

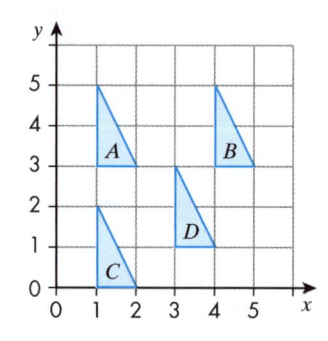

Reflection

A reflection transforms the object to give a mirror image. Each corresponding point in the object and the image are the same perpendicular distance from the **mirror line**, but on opposite sides of it.

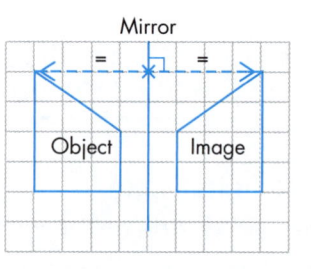

Extended

Shapes may be reflected in any line, not just horizontal or vertical lines.

To find the image of each vertex in the object, draw lines perpendicular to the mirror, and continue them through it, as in the diagram. The **vertices** of the image are the same distance from the mirror line as the corresponding vertices in the object.

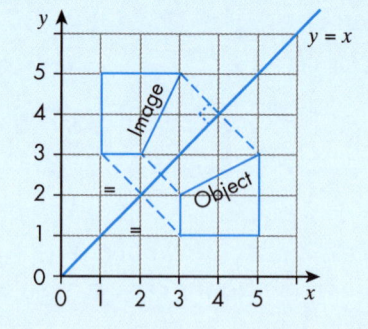

> **Revision tip**
>
> When a shape is reflected in a line that is not vertical or horizontal it is helpful to turn the page around so that the mirror line is vertical or horizontal.

Rotations

A rotation transforms a shape to a new position by turning it about a point called the **centre of rotation**.

To complete a rotation, you require three pieces of information: the centre of rotation, the direction of rotation, the degrees of rotation (usually in multiples of 90°). Note that a rotation of 180° does not need a direction.

> **Revision tip**
>
> Use tracing paper to check that you have made the rotation correctly.

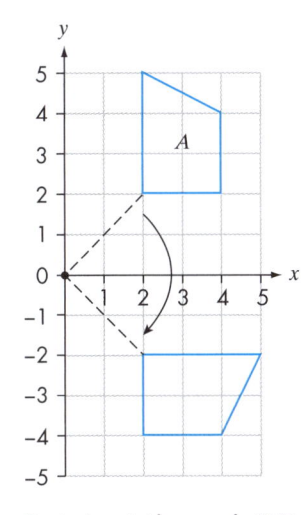

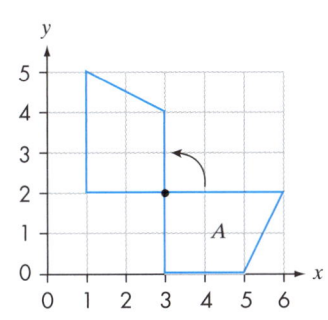

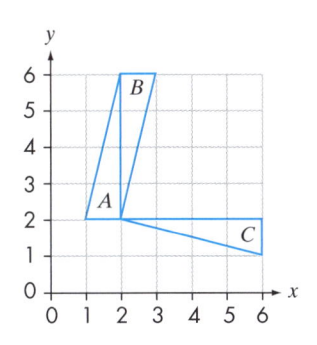

Rotate A through 90° clockwise about the origin (0, 0).

Rotate A through 90° anticlockwise about (3, 2).

In this diagram, what rotation takes triangle A to position B? It is a rotation 180° about (2, 4)
What rotation takes triangle B to position C? It is a rotation 90° clockwise about (2, 2)

Extended

Triangle A has vertices at (1, 3), (2, 3) and (1, 6).

A is rotated through 90° clockwise about (2, 2) to give image B.

B is rotated through 180° about (4, 1) to give image C.

C is rotated through 90° anticlockwise about (5, −2) to give image D.

What rotation will take D to A?

A rotation of 180° about (2.5, 0.5)

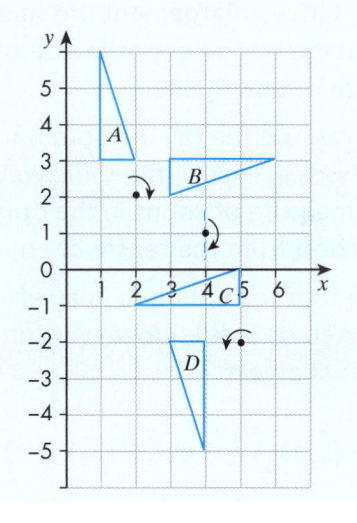

Enlargement

An enlargement changes the size of a shape to give a mathematically similar image – the angles in the shape remain the same but the image is enlarged by a **scale factor**. Enlargements are made from a **centre of enlargement**.
Every length in the enlarged shape is: original length × scale factor
The distance of each image point from the centre of enlargement, O, is: distance of original point from O × scale factor
There are two ways to create the enlargement: the ray method and the coordinate method.
You must use the ray method when the diagram is not on a grid.

Geometry 119

To enlarge triangle ABC by a scale factor of 2, using the ray method

Draw construction lines from the centre of enlargement (COE), O, through each vertex in the diagram and beyond.

Measure the distance from O to A and multiply by the factor of enlargement, 2. This gives the distance along the ray from O to the position of the image vertex, A'.

Repeat to find the positions of B' and C'.

The length of each side of $A'B'C'$ should be twice the length of the corresponding side of ABC.

To enlarge triangle ABC by a factor of 2 from COE (1, 5), using the coordinate method

This method involves counting squares on the grid to work out the new coordinates of the vertices.

First find the distance of A from (1, 5); it is 2 down and 1 across.

Multiply by the scale factor of 2 to get 4 down and 2 across.

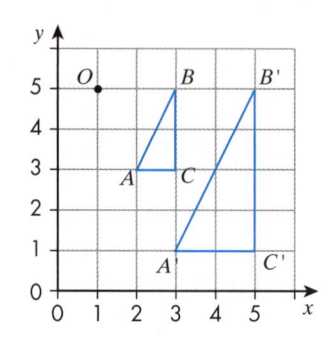

Use these 'measurements' from O to find the position of A'.

Repeat to find the positions of B' and C'.

The length of each side of $A'B'C'$ should be twice the length of the corresponding side of ABC.

A negative enlargement produces an image on the opposite side of the centre of enlargement.

You can use the ray or coordinate methods in exactly the same way, but you measure or count in the opposite direction from the centre of enlargement.

Note that the enlarged image is an inverted or upside-down version of the original object.

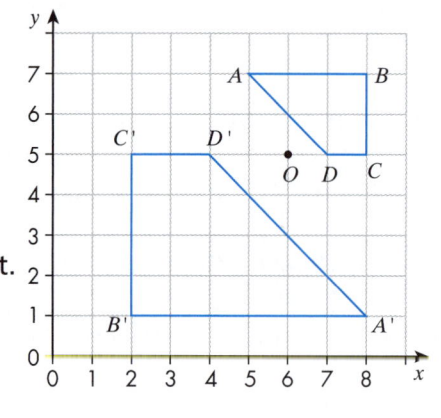

Finding the centre of enlargement

Suppose you know that $A'B'C'D'$ is an enlargement of $ABCD$. You need to be able to find the centre of enlargement and the scale factor of enlargement.

Draw rays joining corresponding vertices in the object and image. They should meet at a single point.

This is the centre of enlargement, in this case, (1, 7).

Measure two corresponding sides to find the factor of enlargement. In this case, $\frac{A'B'}{AB} = \frac{4}{2} = 2$.

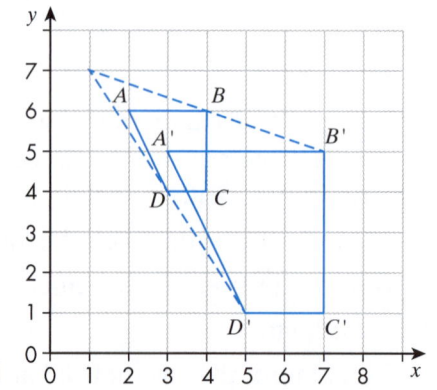

> **Revision tip**
>
> You could draw in the rays to check that the new vertices are in the right positions

120 IGCSE Mathematics Revision Guide

Fractional enlargements

A scale factor may be a fraction. If the fraction is less than 1, the resulting image is smaller than the original, for example, enlarging by a factor of $\frac{1}{2}$ will give an image that is half the size of the original object.

Again, you can use the ray method or the coordinate method.

In the diagram, $ABCD$ is enlarged by a scale factor of $\frac{1}{2}$ from O, so all the distances are multiplied by $\frac{1}{2}$, or the coordinate distances are halved.

The resulting image is $A'B'C'D'$.

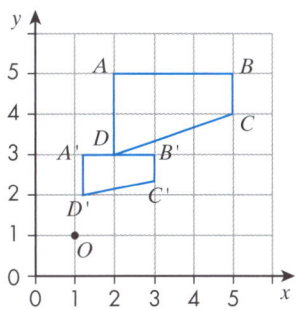

Combined transformations

Sometimes you may be given a shape and then asked to carry out multiple, or combined, transformations on it.

Example

a) Translate A, using the vector $\begin{pmatrix} -2 \\ -2 \end{pmatrix}$, to A_1, and then rotate A_1 through 90° clockwise about $(1, 0)$, to A_2.

What rotation will take A_2 to A?

You need to rotate A_2 through 90° anticlockwise about $(1, 2)$.

b) Translate A, using the vector $\begin{pmatrix} 6 \\ -2 \end{pmatrix}$, to B and then reflect B in the line $y = 1$, to C. Enlarge C, by a scale factor of -2 from $(7, 0)$, to D.

What single transformation will take A to C?

A rotation through 180° about $(6, 2)$.

What single transformation will transform shape A into shape D?

An enlargement by a scale factor of 2 from $(3, 8)$.

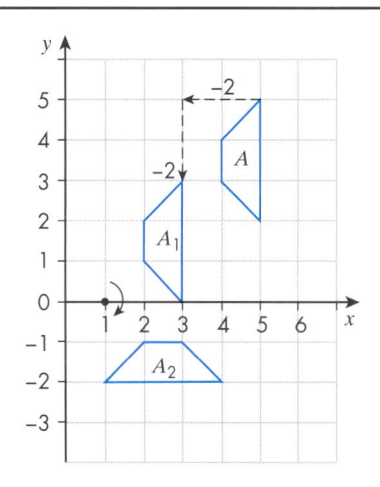

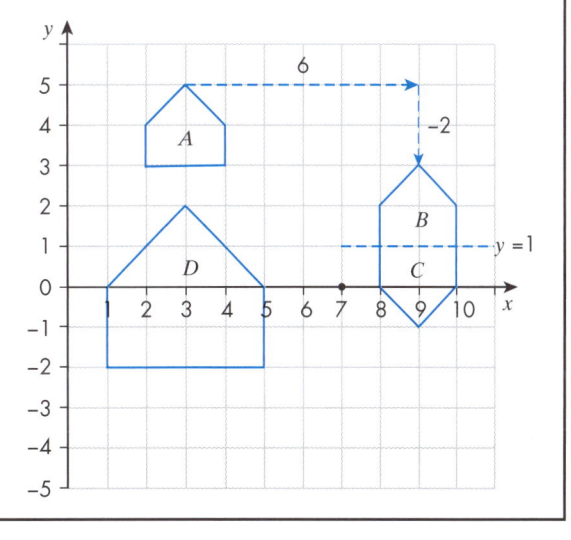

Geometry

Quick test

1. a) Translate shape A using the vector $\begin{pmatrix} -3 \\ -2 \end{pmatrix}$, to give B.

   b) Reflect the shape A in the line $x = 3$, to give C.

Extended

   c) Reflect the shape A in the line $y = x$, to give D.

   d) Rotate the shape A through 90° clockwise about (4, 3), to give E.

   e) Rotate the shape A through 180° about (3, 3), to give F.

   f) What transformation would take C to F?

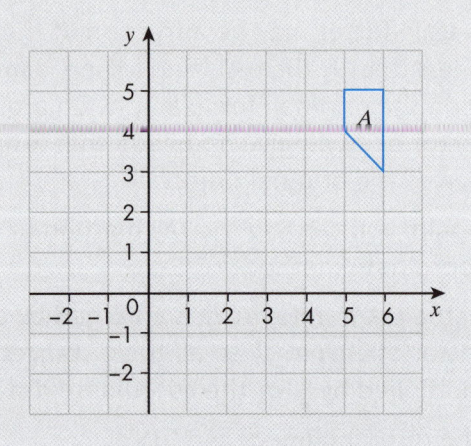

Extended

   g) What transformation would take D to F?

   h) What transformation would take E to F?

2. a) Enlarge shape A by a scale factor of 3 from (0, 1), to give B.

Extended

   b) Enlarge shape A by a scale factor of –1 from (5, 2), to give C.

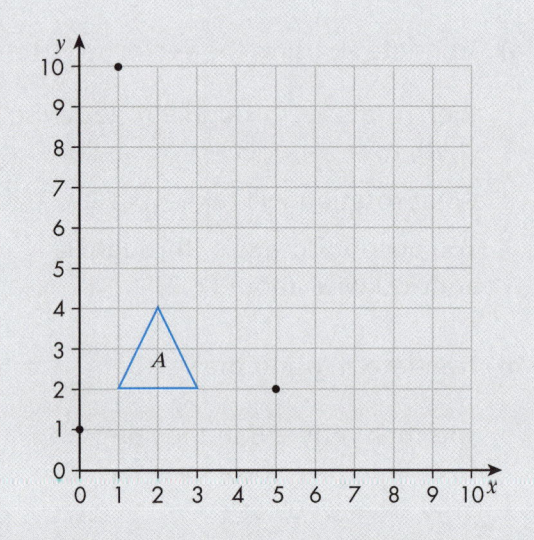

   c) Enlarge shape B by a scale factor of $\frac{1}{2}$ from (1, 10), to give D.

3. a) B is an enlargement of A. What is the scale factor of enlargement and where is the centre of enlargement?

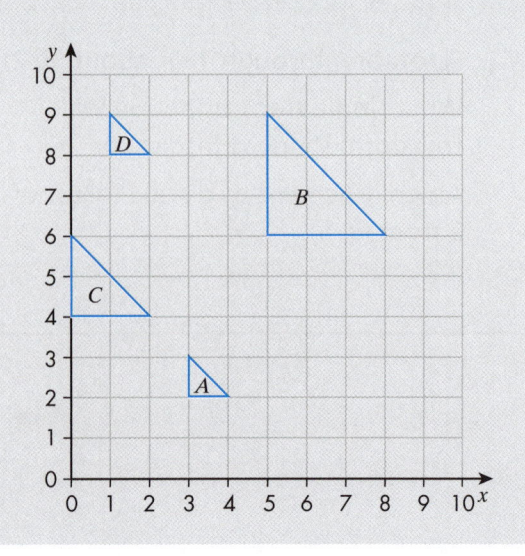

   b) C is an enlargement of A. What is the scale factor of enlargement and where is the centre of enlargement?

Extended

   c) D is an enlargement of B. What is the scale factor of enlargement and where is the centre of enlargement?

Exam-style practice questions

1
a) **i)** Calculate the value of x and the size of angle $(2x + 7)$.

ii) Calculate the value of y and the size of angle $(3y - 5)$.

iii) Using your answers to part i) and ii) work out the value of z. [3]

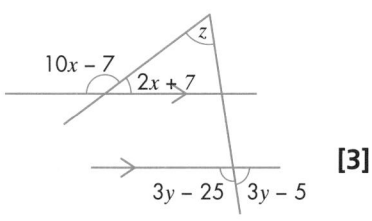

Extended

b) Majida says that this shape is a parallelogram. Natasha says it is a trapezium.
Who is right? Why? [3]

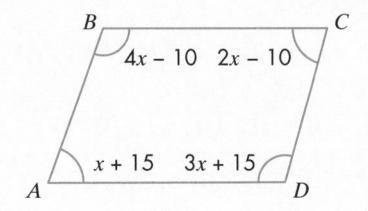

c) Tony and Kim have measured and totalled the interior angles of the same regular polygon. Tony has $1880°$ and Kim has $1980°$. Who is correct? How many sides does the polygon have? [3]

d) AB is a tangent to the circle, centre O.

Calculate the value of x. [2]

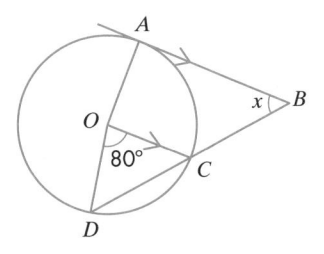

Extended

e) O is the centre of the circle. Calculate the values of x and y. [5]

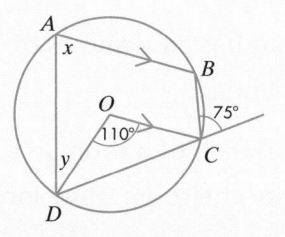

f) O is the centre of the circle. EBD and FCD are tangents to the circle. Calculate the values of a and b.

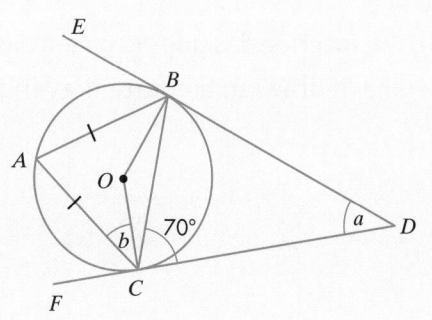

2 **a)** Use a ruler and protractor to draw this triangle accurately. What is the size of angle C and the length of side AC? [3]

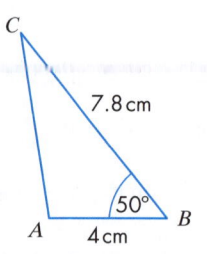

b) A marker buoy is 10 km due south of a lighthouse and a ship is 6 km from the buoy on a bearing of 065°. Make a scale drawing of this and measure the bearing of the lighthouse from the ship. [3]

c) Will this net fold into a tetrahedron? [1]

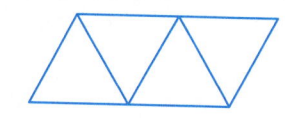

d) Are these two triangles similar? Justify your answer. [2]

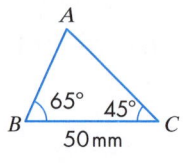

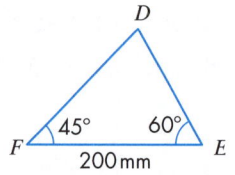

Extended

e) The design sheet for a new banknote has a length of 100 cm and an area of 4500 cm². The actual banknote will have an area of 76.05 cm². What will the length of the actual banknote be? [3]

f) Fruit juice is sold in two similar cuboid cartons: snack and family. The snack carton has a base area of 18 cm², the area of the base of the family carton is 60 cm². If the snack carton holds 180 ml of juice, what is the capacity of the family carton (to 2 s.f.)? [3]

g) A kite is a special quadrilateral. Draw a picture of a kite and include its lines of symmetry. Are the two triangles that make up the kite congruent? Explain your answer. [2]

3 **a)** Construct an isosceles triangle that has base angles of 70° and two sides of 6 cm. What is the length of the third side of the triangle? [3]

b) An interior designer is given this sketch of a room in a house. Using a scale of 2 cm = 1 m, make draw an accurate drawing of the room. How long is wall CD? [3]

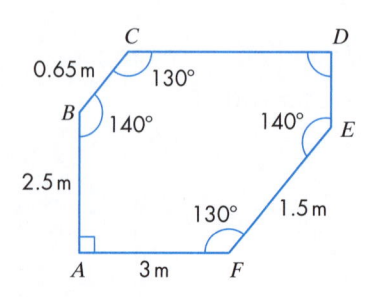

4

a) Calculate the length x, to 3 s.f. [2]

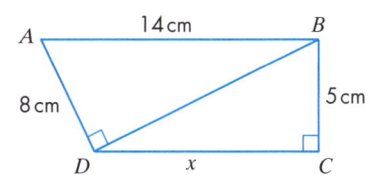

b) Calculate the length x, to 3 s.f. [2]

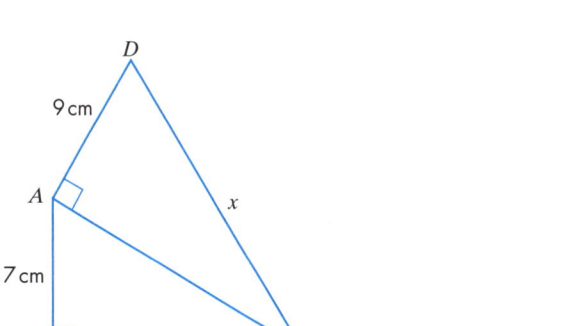

c) Using the diagram in **Q4a)**, calculate Angle ABD to 1 d.p. [3]

d) In triangle ABC, Angle $A = 90°$, $BC = 16$ cm and $AC = 14$ cm. What is the size of Angle C to 1 d.p.? [2]

e) Using the diagram in **Q4b)**, calculate Angle BAC to 1 d.p. [2]

Extended

f) Rashidi, who is 1.7 m tall, is standing in Trafalgar Square, London, looking up to the top of Nelson's Column, which has a height of 52 m. The angle of elevation is 71°. How far away from the base of the column is Rashidi standing, to nearest m? [2]

g) The square-based pyramid $ABCDE$ has a vertical height, EF, of 15 cm. $AD = 12$ cm. Calculate Angle AEC to 1 d.p. [4]

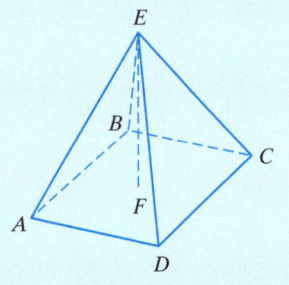

h) In Figure 1, if Angle $BAC = 55°$, what is the size of Angle ABC to 1 d.p.? [4]

i) In Figure 1, if Angle $BCA = 47°$, what is the length of AB (to 2d.p.)? [3]

j) In Figure 1, if Angle $BCA = 47°$, what is the area of the triangle (to 2 d.p.)? [2]

Figure 1

Geometry 125

5 **a)** This equilateral triangle has side lengths of 10 cm. What is the area of the rectangle that surrounds it (to 2 d.p.)? [3]

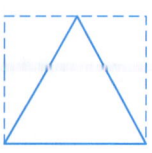

b) A square pond of side 3 m is being reshaped into a circle. The surface area of the water is to remain the same. What is the diameter and circumference of the new circular pond to 2 d.p.? [3]

c) A children's toy comes in a small cuboid shaped box with sides 20 cm, 10 cm, 8 cm. They are packed into delivery crates that are cubes of side length 0.8 m. How many toys will fit into 1 crate? [4]

d) A small metal component in a toy car is a cylinder of radius 1 cm and length 4 cm. A square hole of sides 6 mm is punched down its length. [4]

  i) What is the volume of metal in this component (to 2 d.p.)? [4]

  ii) The component is made from an aluminium alloy with a density of 2.72 g/cm^3. What is the mass of the component, to the nearest g? [2]

e) A jelly mould has the dimensions in the diagram. How much liquid jelly can be poured into the mould? Answer in millilitres, to the nearest millilitre. [3]

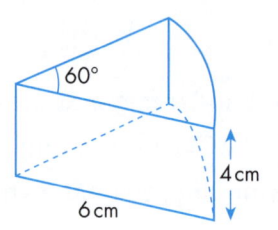

Extended

f) A paper weight is made to the dimensions shown. It is made from flint glass which has a density of 3.5g/cm³. What is the mass of the paper-weight (to 2 s.f.)? **[4]**

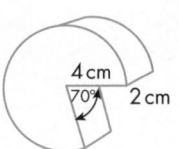

g) A children's toy is in the shape of a cone with a hemispherical base as shown.

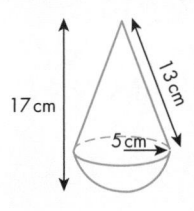

  i) What is its volume? **[4]**

  ii) What is its total surface area, leave your answer in terms of π? **[2]**

6 a) The diagram shows a drawing of the top of a diamond. How many lines of symmetry does it have? **[1]**

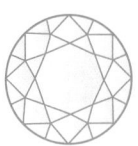

Extended

b) In the diagram AC and BC are tangents to the circle, centre O.

  i) What shape is $OACB$? **[1]**

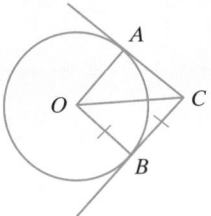

  ii) What is the size of Angle AOC? **[2]**

7 a) M has coordinates (1, 3), N is at (5, 5) and S is at (3, −1). P is the midpoint of MN, Q the midpoint of NS and R the midpoint of MO. Mark these points on a coordinate grid.

Write down the column vectors $\overrightarrow{MQ}, \overrightarrow{RN}, \overrightarrow{SP}, \overrightarrow{NS}, \overrightarrow{QR}$ and $\overrightarrow{QP}$. **[6]**

Extended

b) In the diagram, S is the midpoint of BD.

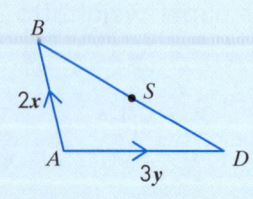

i) Work out the vector $\overrightarrow{BD}$. [2]

ii) Work out the vector $\overrightarrow{AS}$. [2]

The line AS is extended so that $SC = AS$.

iii) Show that BC is parallel to AD. [2]

c) $\mathbf{A} = \begin{pmatrix} 1 \\ -2 \end{pmatrix}$, $\mathbf{B} = \begin{pmatrix} 4 \\ 4 \end{pmatrix}$, $\mathbf{C} = \begin{pmatrix} 2 \\ 5 \end{pmatrix}$

i) Calculate |**AB**| and |**BC**|, leaving your answers in surd form. [3]

ii) Abeje says that Angle $ABC = 90°$. Is this true? Justify your answer. [2]

8 a) Make a copy of the diagram.

Use your diagram to answer the following questions.

i) Shape X is translated through the vector $\begin{pmatrix} -3 \\ 1 \end{pmatrix}$ to produce image A. Where was shape X? [2]

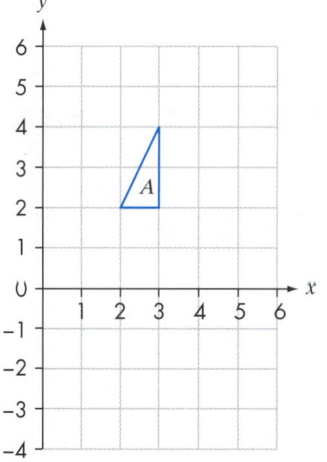

ii) Translate shape A through the vector $\begin{pmatrix} 2 \\ 2 \end{pmatrix}$ to image B. [2]

iii) Reflect shape A in the mirror line $x = 2$ to produce image C. [2]

iv) Reflect shape A in the x-axis to produce image D. [2]

Extended

b) Use this diagram for these questions.

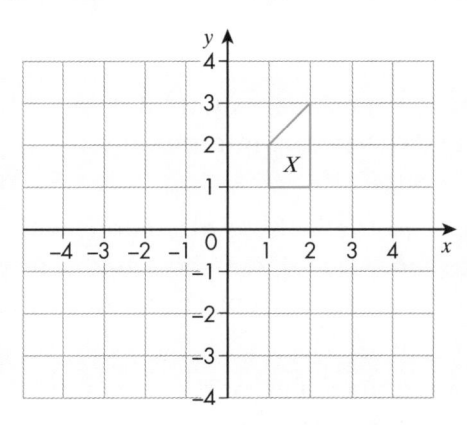

i) Reflect shape X in the line $y = -x$, to produce image A. [2]

ii) Reflect shape X in the line $y = x + 1$, to produce image B. [2]

c) Use the diagram in **Q8a**. Rotate shape A through 90° anticlockwise about (4, 0), to produce image B. [3]

d) Use the diagram in **Q8b**.

  i) Enlarge shape X by a scale factor of 2, centre of enlargement (5, 2), to produce image A. [3]

  ii) Enlarge shape X by a scale factor of -2, centre of enlargement (-1, 0), to produce image B. [3]

  iii) What rotation would take shape A (from part **i**) to shape B (from part **ii**)? [2]

e) Enlarge shape X by a scale factor of $\frac{1}{2}$, centre of enlargement (8, -2), to produce image A. [3]

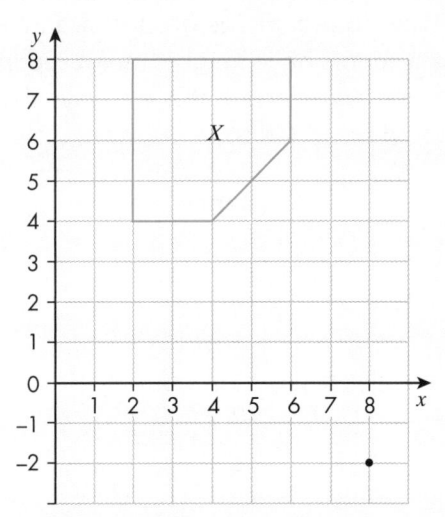

Geometry 129

Statistical representation

Frequency tables

Statistics is concerned with the collection and organisation of data, representing that data in diagrams and charts, and interpreting the data and diagrams.

When data is collected, it needs to be recorded. This is often done with a tally chart (usually leading to a **frequency table**). A mark is made as each instance of the data is recorded. These markers are counted to give a **frequency**.

For example, Kate counted and recorded the makes of the cars in the school car park and presented her results in a frequency table.

Make	Tally	Frequency
Make A	⦀⦀ ⦀⦀ ⦀⦀ ⦀⦀ ///	23
Make B	⦀⦀ ⦀⦀	10
Make C	⦀⦀ ////	9
Make D	⦀⦀ ⦀⦀ ⦀⦀	15
Other	⦀⦀ ⦀⦀ ///	13

Some data, such as mass, height and test scores, may take a wide range of values. When recording heights of a large number of people, it would not be sensible to count how many people were 140 cm tall or 156 cm tall. In such cases the data is grouped.

This frequency table represents the heights of students in Kate's tutor group. Height is continuous data (because the data can take an infinite number of values on a scale) and is always rounded. The groups of data are called classes or class intervals, so nine students have heights measured in the range $140 \leq h < 150$ cm tall. This notation means that it is always clear which class interval a piece of data should be added to.

Height (cm)	Tally	Frequency
$130 \leq h < 140$	//	2
$140 \leq h < 150$	⦀⦀ ////	9
$150 \leq h < 160$	⦀⦀ ⦀⦀ ⦀⦀ ////	19
$160 \leq h < 170$	⦀⦀ ⦀⦀ //	12
$170 \leq h < 180$	///	3

Pictograms

The results of a survey may be presented in pictorial or diagrammatic form. A **pictogram** is a frequency diagram with symbols representing frequency. A pictogram needs a key indicating how many items are represented by one symbol. Pictograms are very easy to understand, although sometimes fractions of symbols may be difficult to draw accurately.

> **Revision tip**
>
> Make sure the symbol you use is easy to draw accurately and understand. Use tracing paper to help you to draw an accurate pictogram if you are given a partially completed pictogram in a question.

Here is the information about cars in the car park drawn as a pictogram.

Make	🚗 = 4 cars
Make A	🚗🚗🚗🚗🚗🚗
Make B	🚗🚗🚗
Make C	🚗🚗🚗
Make D	🚗🚗🚗🚗
Other	🚗🚗🚗🚗

Bar charts

A **bar chart** is made up of a series of bars, of the same width. These may be drawn vertically or horizontally, from an axis that indicates the categories of data the bars represent. The height or length of each bar represents the frequency of that data item. Small gaps are left between the bars. Gaps usually need to be all the same width and are there because bar charts are used to represent categorical data or discrete data.

This bar chart records which day's menu the students in Kate's tutor group preferred.

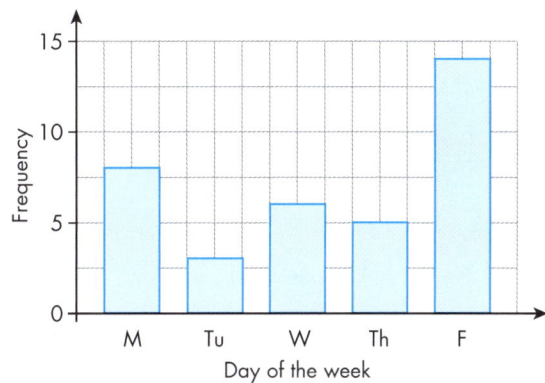

Dual bar charts record information about two sets of related data, making comparisons easier. This one show how boys and girls get to school.

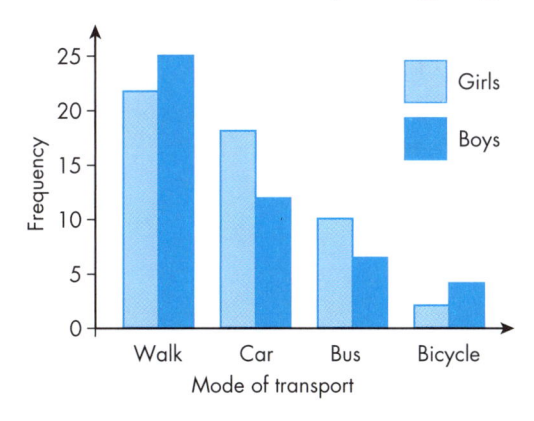

> **Revision tip**
>
> Give a key for dual bar charts.

Pie charts

In a **pie chart** a circle (the 'pie') represents all the data. Each category of data is represented by a sector (a 'slice'). The angle of the sector is proportional to the frequency of the category it represents, meaning that one disadvantage of the pie chart is that it cannot show individual frequencies.

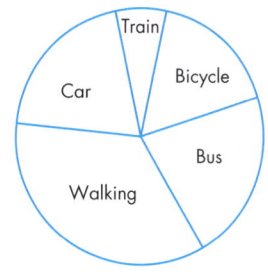

This pie chart shows the proportions of methods of transport that a total of 60 students use to travel to school.

Statistics 131

The original data was:

Transport	Bicycle	Bus	Walking	Car	Train
Students	10	13	21	12	4

To work out the angle for each sector find the fraction of 360° that represents each category.

For example:

10 out of the 60 travel by bicycle, so their angle is $\frac{10}{60} \times 360° = 60°$ and the angle for students travelling by bus is $\frac{13}{60} \times 360° = 78°$

> **Revision tip**
>
> Remember to add up all the sector angles to make sure they total 360°.

Stem and leaf diagrams

Another way of presenting data is in a stem and leaf diagram.

For example: the total number of medals won by the Great British team at the Olympic Games since 1928: 20, 16, 14, 23, 11, 24, 20, 18, 13, 18, 13, 21, 37, 24, 20, 15, 28, 30, 48, 65, 67

First, rearrange this data making the first digit of the medal number the 'stem' and making the second digit a 'leaf':

1	6 4 1 8 3 8 3 5
2	0 3 4 0 1 4 0 8
3	7 0
4	8
5	
6	5 7

Stem and leaf diagrams need to be ordered so the second step is to put the leaves of each row into ascending order.

A key needs to be added so that someone reading the diagram can interpret it correctly.

1	1 3 3 4 5 6 8 8
2	0 0 0 1 3 4 4 8
3	0 7
4	8
5	
6	5 7 key 1	1 means 11

These diagrams can be used to identify the mode and median of the data set (see the next section 'Statistical measures' for more detail on these).

The median is the middle value (11th out of 21) so counting left to right starting on the top line find that the median is 20 medals.

The mode is the value that occurs most frequently which can be seen to be 20 medals.

Scatter diagrams

Scatter diagrams, or **scattergraphs**, are created by plotting two sets of variable data against each other as (x, y) coordinates.

The plotted points form one of three patterns, showing three possible types of correlation.

- **Positive correlation** means that as one quantity increases so does the other – for example, as temperature rises, the number of ice-creams sold also increases.

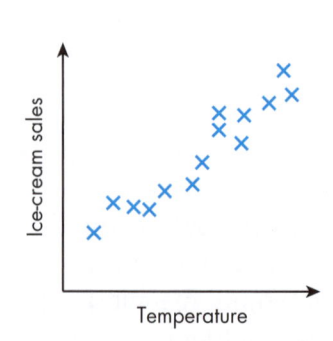

IGCSE Mathematics Revision Guide

- **Negative correlation** means that as one quantity increases the other decreases – for example, as the number of people vaccinated against a disease increases, fewer people contract the disease.

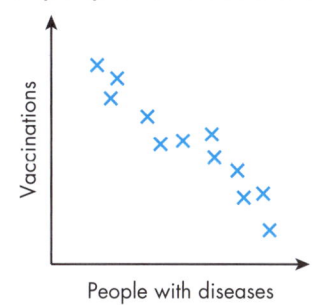

- **No correlation** or zero correlation means that there is no recognisable connection between the two data sets – for example, the length of your hair is no guide to your IQ level.

For positive or negative correlation, you can draw a **line of best fit**, which is a straight line that passes between, and as close as possible to, all points in the scatter diagram. There should be roughly the same number of points on each side of the line.

The line of best fit for the ice-cream–temperature chart would look like this.

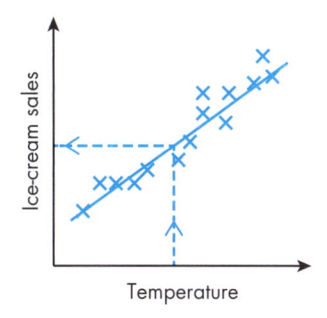

You can use the line of best fit to estimate the answer to questions such as, 'How many ice-creams will be sold if the temperature is…? Draw a line from that temperature on the 'Temperature' axis up to the line of best fit and then across to the 'Ice cream sales' axis. This gives an estimate of the sales for that temperature.

Histograms

Histograms look very similar to bar charts, but there are significant differences.

- There are no gaps between the bars.
- The horizontal axis has a continuous scale.
- The vertical scale is labelled 'frequency density' as area represents frequency (see *Histograms with bars of unequal width*, page 134).

The histogram below has been drawn from the table, which records how long it took a group of people to complete a jigsaw.

Time, h, hours	$0 < h \leqslant 2$	$2 < h \leqslant 4$	$4 < h \leqslant 6$	$6 < h \leqslant 8$
Frequency	3	8	12	7

The columns are not individually labelled; the horizontal axis has a continuous scale. The columns are drawn to represent the intervals in the table, $0 < h \leqslant 2$, $2 < h \leqslant 4$ and so on.

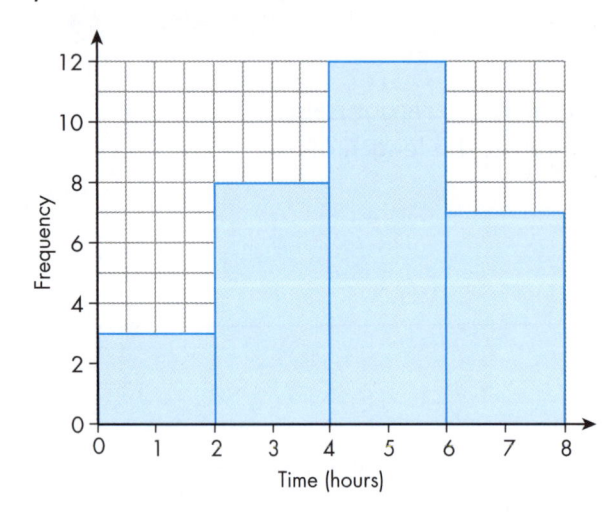

Extended

Histograms with classes of unequal width

If a set of data has class intervals of different widths, the rectangles of the histogram will also have different widths. In this case, the area of the bar, not its height, represents its frequency. You find the height of the bar by dividing the class frequency by its class-width (its width, which is found by subtracting the interval's lower bound from its upper bound). This measure is the **frequency density** and is the label used on the vertical axis.

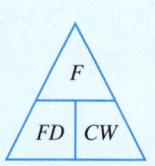

frequency density, FD = frequency ÷ class width = $F ÷ CW$

For example, this table shows the number of hours a day that 100 students spent on social media over a holiday period.

Hours, h	$0 < h \leqslant 2$	$2 < h \leqslant 4$	$4 < h \leqslant 5$	$5 < h \leqslant 6$	$6 < h \leqslant 7$	$7 < h \leqslant 10$
Frequency	16	20	15	18	16	15

You need to calculate the frequency density for each class interval. Redraw the table vertically and add some columns. The first few rows will look like this.

Hours, h	Frequency	Class-width	Frequency density
$0 < h \leqslant 2$	16	2	8
$2 < h \leqslant 4$	20	2	10
$4 < h \leqslant 5$	15	1	15

Then you can draw the histogram, like this:

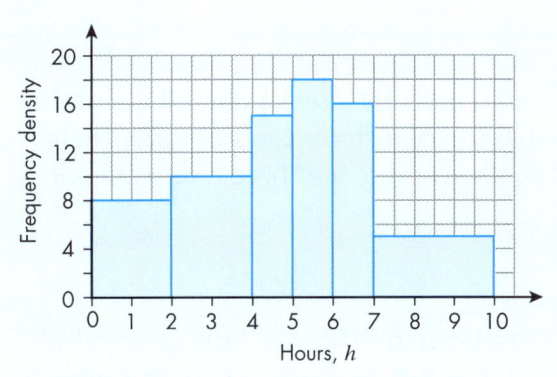

Quick test

1. A group of teenagers was asked to vote for their favourite pop artist. Here are the results.

Artist	Olly Murs	Little Mix	Lukas Graham	Zayn	Jonas Blue	Shawn Mendes
Votes	8	20	14	12	5	1

From this information draw: **a)** a bar chart **b)** a pie chart.

2. Every day, for a school year, Amar counted the people who were on his bus when he got on in the morning. Here are the results. Draw a histogram from this information.

Passengers, p	$0 < p \leq 10$	$10 < p \leq 20$	$20 < p \leq 30$	$30 < p \leq 40$	$40 < p \leq 50$	$50 < p \leq 60$
Days	3	5	10	32	40	25

3. The number of people taking a ferry between an island and the mainland on each morning of 25 working days has been counted. These are the results:

   42 63 95 74 86 68 76 88 89 41 58 91 53
   56 90 94 71 81 73 84 54 69 68 82 47

   Place and arrange these values into an ordered stem and leaf diagram and identify the median and mode number of passengers.

Extended

4. 150 students were asked to find a character hidden in a picture. Their times, in seconds, were recorded. Here are the results. Draw a histogram from this information.

Time, s	$0 < s \leq 10$	$10 < s \leq 15$	$15 < s \leq 20$	$20 < s \leq 25$	$25 < s \leq 35$	$35 < s \leq 60$
Students	19	24	32	29	26	20

Statistical measures

The term **average** is commonly used to describe a set of data. It is a useful 'typical' value that allows for comparison of different sets of data. There are a number of ways of expressing an average, the three most common being: **mode**, **median** and **mean**.

The mode

The mode is the value that occurs most times in a data set – the value with the highest frequency. In a bar chart this would be the data item that had the tallest bar. The mode is a useful average because it can be used when describing non-numerical data, such as favourite items on a restaurant menu.

Suppose a farmer records the number of eggs laid by his chickens for each of 15 days.

    9 8 6 7 4 7 6 8 6 9 5 5 8 9 8

The number that occurs most often (most frequently) is 8, so the mode is 8.

Another way to describe this is to say that the **modal value** is 8. For grouped data the modal group or modal class is the class which has the greatest frequency.

The median

The median is the middle value in a list of data when they have been sorted in order of size. This means that half of the data values are above the median and half of them are below. The median value is not particularly affected by any extreme low or high values.

Sorting the number of eggs that the farmer's chicken have laid gives this list.

    4 5 5 6 6 6 7 7 8 8 8 8 9 9 9

There are 15 days so the median is the 8th value, 7 eggs.

If the farmer had counted for 16 days, and on the 16th day he found 10 eggs, then the list would look like this.

    4 5 5 6 6 6 7 7 8 8 8 8 9 9 9 10

This means that the median is between the 8th and 9th values. Then the median is $(7 + 8) \div 2 = 7.5$ eggs.

> **Revision tip**
>
> Sorting the data into order helps you to identify the mode as well as the median.

Extended

If you are presenting the data in a frequency table, add a new row so you can add up the frequencies to give a running total, called the **cumulative frequency**.

The farmer counts the number of eggs for 100 days:

Eggs	4	5	6	7	8	9	10
Number of days	9	8	22	17	15	16	13
Cumulative frequency	9	17	39	56	71	87	100

The median is between the 50th and 51st values. The first 39 values include all the days 6 eggs or fewer were found, so the values for 7 eggs start at the 40th and go up to and include the 56th. So the median number of eggs is 7.

The mean

The mean is what most people understand by the term 'average'.

$$\text{mean} = \frac{\text{sum of all values}}{\text{total number of values}}$$

The advantage of the mean is that it uses all values in the data set.

For example, 10 students counted the number of books they each had in their school locker.

6   4   5   7   6   4   11   7   9   7

The mean number of books per student = $\frac{66}{10}$ = 6.6

Because the mean involves a division it may not be an integer, and often isn't a possible value from the data set.

The range

The **range** of a set of values is the difference between the largest and smallest values. It is not an average; it shows the spread of the data. It is useful when comparing sets of data because it can help you comment on the consistency of data.

For example, Ben and Rav each play 10 innings of cricket. These are their scores.

| **Ben** | 27 | 34 | 36 | 11 | 23 | 29 | 37 | 59 | 19 | 45 |
| **Rav** | 29 | 26 | 38 | 25 | 30 | 36 | 39 | 35 | 29 | 33 |

Both have a mean score of 32. Ben's range is 48 and Rav's is 14, making Rav the more consistent batsman.

Which average to use

The average that you choose to use for a data set must be appropriate and representative of the data. If you do not use the most appropriate average, your results and conclusions could be misleading.

	Mode	**Median**	**Mean**
Advantages	Very easy to find Not affected by extreme values Can be used for non-numerical data	Easy to find for ungrouped data Not affected by extreme values	Easy to find Uses all the values The total for a given number of values can be calculated from it
Disadvantages	Does not use all the values May not exist or may be more than one mode	Does not use all the values Often not understood	Extreme values can distort it Has to be calculated
Use for	Non-numerical data Finding the most likely value	Data with extreme values	Data with values that are spread in a balanced way

Look at this data set.

6 10 4 5 4 3 3 10 2 3

The mode is 3, the median is 4 and the mean is 5. Depending on what the data set represents, one of these averages might be more appropriate than the others:

- If you were coaching people for sporting competitions you could say that the average ranking achieved was 3 (mode).
- In one hour the average shoe size of customers in a shoe shop was size 4 (median).
- The average mark in a quick spelling test was 5 (mean).

> **Revision tip**
>
> Think carefully about the data set and its context when you need to select the most appropriate average to use.

These measures can be used to compare sets of data. For example, Jim and Dmitri are throwing a tennis ball at a tin can on a wall, trying to knock it down. They record the number of throws they take to hit the can off the wall. They play the game 40 times each. The results are in this table:

Number of throws	1	2	3	4	5	6	7	8	9	10
Jim	0	3	5	10	8	7	5	2	0	0
Dmitri	1	5	6	8	10	4	3	1	1	1

Comparing the statistical measures gives:

	Jim	Dmitri
Mean	4.85	4.875
Mode	4	5
Median	5	4.5
Range	6	9

The mean values suggest that Dmitri knocks slightly more cans off the wall, as does the mode, however, the median suggests that Jim knocks more cans off the wall. Jim has a smaller range, suggesting he is more consistent. Using the range can sometimes be unhelpful because it only takes one extreme data item to distort the comparison, though this is not the case here.

Extended

Using frequency tables

Rather than listing every value recorded in a large collection of data, it is often more convenient to use a frequency table. This table records the number of children living in 100 households.

Number of children (x)	1	2	3	4	5	6
Frequency (f)	16	31	15	17	12	9

From this table, you can find all three averages, and the range.

To find the median, add a new row to work out the cumulative frequency.

To find the mean, add another new row to find the value of the number of children, x, multiplied by the frequency, f, to give fx.

The total of all the frequencies is written as Σf and the total of all values of fx is Σfx.

Number of children (x)	1	2	3	4	5	6	Total
Frequency (f)	16	31	15	17	12	9	100
Cumulative frequency	16	47	62	79	91	100	
fx	16	62	45	68	60	54	305

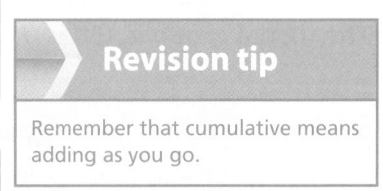

> **Revision tip**
> Remember that cumulative means adding as you go.

The mode is 2 children per household (highest frequency, 31 households).

The median is between the 50th and 51st values. The cumulative frequency indicates that this is 3 children per household (3 children per household starts at the 48th value and continues up to and including the 62nd value).

The mean is found as: $\frac{\Sigma fx}{\Sigma f} = \frac{305}{100} = 3.05$ children per household.

Grouped data

The data considered in the examples so far in this section has been **discrete data**, which is data that consists of separate values, such as number of goals scored or number of children.

Normally, grouped tables use **continuous data**, which is data that can have any value in a given range, such as height, mass or speed.

You can calculate the mean of grouped data, but in this case it is an estimate because the grouped data summarises the original data. You can also identify the **modal group.**

Anna kept a record of the miles she travelled each day over a period of 100 days in her job as a coach driver.

Mileage (d)	f
0 ⩽ d < 50	3
50 ⩽ d < 100	11
100 ⩽ d < 150	23
150 ⩽ d < 200	30
200 ⩽ d < 250	19
250 ⩽ d < 300	9
300 ⩽ d < 350	5

The problem here is that you have no specific value to multiply with f because the values for d are grouped. Therefore, you find the midpoint of each class, and use that as the value to multiply by f.

Milage (d)	f	Midpoint of d (x)	fx
0 ⩽ d < 50	3	25	75
50 ⩽ d < 100	11	75	825
100 ⩽ d < 150	23	125	2875
150 ⩽ d < 200	30	175	5250
200 ⩽ d < 250	19	225	4275
250 ⩽ d < 300	9	275	2475
300 ⩽ d < 350	5	325	1625
	Σf = 100		Σfx = 17400

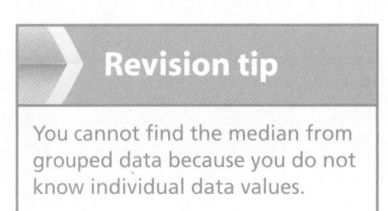

> **Revision tip**
> You cannot find the median from grouped data because you do not know individual data values.

Statistics

The mean is: $\frac{\Sigma fx}{\Sigma f} = \frac{17400}{100} = 174$ miles/day.

The modal group is $150 \leq d < 200$ miles/day.

Cumulative frequency diagrams

A cumulative frequency diagram, or curve, is drawn using the cumulative frequency data as illustrated in the examples on the previous page.

Anna's 100 days of coach driving produces the following cumulative frequencies.

Mileage (d)	f	Cumulative frequency
$0 \leq d < 50$	3	3
$50 \leq d < 100$	11	14
$100 \leq d < 150$	23	37
$150 \leq d < 200$	30	67
$200 \leq d < 250$	19	86
$250 \leq d < 300$	9	95
$300 \leq d < 350$	5	100

The cumulative frequency can be plotted against the mileage. With grouped data, you use the upper bound of the class interval, so the first two points to plot would be (50, 3) and (100, 14). Then join the points with a smooth curve.

The median is the middle value in the sorted data set. There are 100 values in Anna's data so the middle value is at position 50. Draw a line from 50 on the vertical cumulative frequency axis, across to the curve and down to the horizontal mileage axis.

175 miles is the estimated median.

By splitting the cumulative frequency into four parts you can find the **lower quartile** at 25% of the data, position 25 of Anna's data, and the **upper quartile** at 75%, position 75 of the data.

Lower quartile, $Q_1 = 130$ miles

Upper quartile, $Q_3 = 220$ miles

The **interquartile range** (IQR) is $220 - 130 = 90$ miles.

$$IQR = Q_3 - Q_1$$

The IQR, the middle 50% of the data, is the most representative portion and ignores the extreme values.

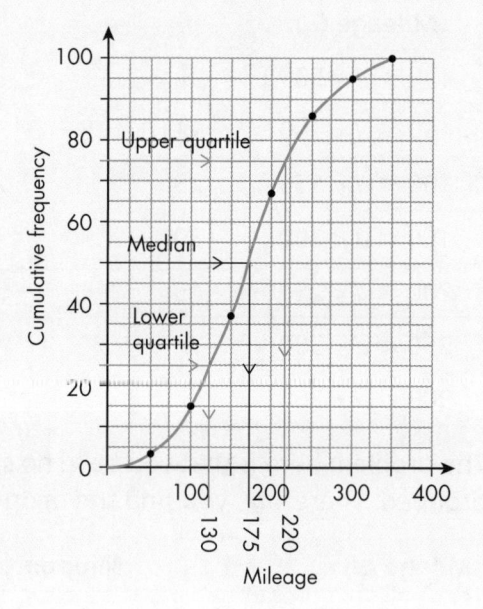

IGCSE Mathematics Revision Guide

The students in a year group are timed to see how long it takes them to complete a puzzle. The results are in this table:

Time (minutes)	$4 < t \leq 6$	$6 < t \leq 8$	$8 < t \leq 10$	$10 < t \leq 12$	$12 < t \leq 14$	$14 < t \leq 16$
Boys	1	5	14	25	9	6
Girls	2	8	23	16	7	4

If cumulative frequency curves are drawn for each of these two sets of data and plotted on the same graph space, you get this graph:

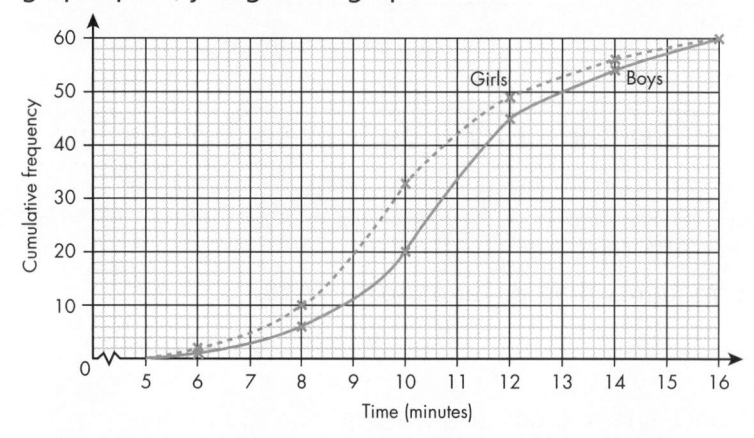

The curve for the girls shows them quicker than the boys at all stages. If the median, lower and upper quartiles were taken you would find that the values for the girls were all smaller than for the boys. The only value not easily seen from the graphs is the interquartile range. In fact, the girls' IQR is 2.8 minutes, but the boys' is 2.4 minutes, which means the boys times are more consistent than the girls times.

Box and whisker diagrams (box plots)

These diagrams are a very useful of displaying and comparing data. They require five summary values the lowest value in the data set, the lower quartile (Q_1), the median (Q_2), the upper quartile (Q_3) and the highest value in the data.

Using Anna's coach driving data from the cumulative frequency diagram on the previous page, the box and whisker diagram would look like this:

Barack works for the same coach company and a box and whisker diagram of his mileage looks like this.

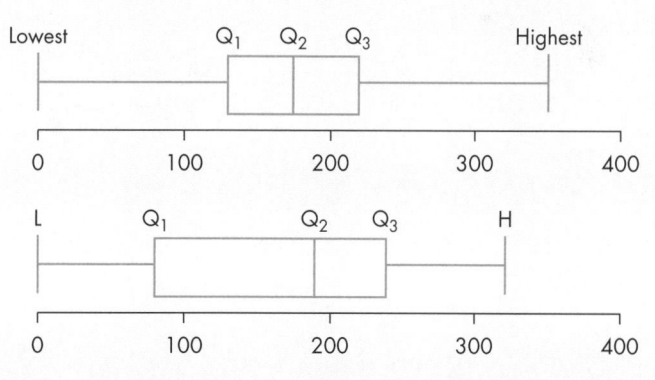

Standard observations could include: Barack's IQR is larger; Anna's range is larger; Barack's median value is higher. Sometimes you will be asked to interpret the information that the diagrams show. For this, you need to convey an understanding of the significance of the elements in the diagrams. For example, Anna has a smaller IQR. This represents 50% of her journeys and could suggest that she does the same journey many times – there is little variation between the majority of distances she travels. Anna does, however, have a larger range than Barack, but it only takes one very long journey to influence this measurement – perhaps she only went on a journey of approximately 350 miles once, but it was still recorded. 25% of Barack's journeys were of distances less than approximately 80 miles, compared to approximately 130 miles for Anna, so he seems to have been used to drive on more shorter journeys than Anna. Both drivers had at least one day where they did not drive – both diagrams have a lowest data recording of 0 miles.

Statistics

Quick test

1. Work out the mode, median, mean (to 2 d.p.) and the range for each of the following data sets.

   a) 29 30 36 40 28 21 34 36 49 24 25 26

Extended

   b)
x	0	1	2	3	4	5	6	7	8	9	10
Frequency	2	5	6	7	11	13	16	14	10	9	5

2. The sycamore tree has the largest leaves of any tree in the UK. The table shows the area, a in cm^2, of fallen leaves collected from one tree in a garden.

Leaf area (a)	$100 \leq a < 150$	$150 \leq a < 200$	$200 \leq a < 250$	$250 \leq a < 300$	$300 \leq a < 350$
Number of leaves (f)	51	67	118	93	71

   Calculate an estimate for the mean area of a sycamore leaf.

3. Using the data from Q2, create a cumulative frequency diagram and use it to find:

   a) an estimate of the median
   b) the interquartile range.

4. Using your cumulative frequency from question 3, create a box and whisker diagram for the sycamore data.

IGCSE Mathematics Revision Guide

Probability

The probability scale

Probability is a measure of chance, or the likelihood that something will happen.

What is the likelihood that it will snow tomorrow? What is the chance that I will win this race?

Impossible | Very unlikely | Unlikely | Even chance | Likely | Very likely | Certain

0 — $\frac{1}{2}$ — 1

Probability is measured on a scale of 0 to 1.

A probability of 0 means that a particular **event** or **outcome**, such as 'the sun will not rise tomorrow', is impossible.

A probability of 1 means that a particular event or outcome, such as 'the day after Wednesday will be Thursday', is certain.

All **events** or **outcomes** have probabilities between 0 and 1

Using the words on the scale, what is the probability that

- you will score an odd number when you throw a dice? *Even chance*
- when you wake up in the morning you will find you have grown 10 cm overnight? *Impossible*
- you will turn 16 years old on your next birthday if you are 15 years old now? *Certain*
- you will be given homework at your next maths lesson? *Likely*

Calculating probabilities

If you are asked to write down the probability of an event happening, you may have difficulty deciding what number to use.

It is possible to calculate a probability by considering two things: the number of ways that the required outcome can happen and the total number of possible outcomes.

These pieces of information can then be written as a fraction.

$$P(\text{outcome}) = \frac{\text{number of ways the required outcome can happen}}{\text{total number of possible outcomes}}$$

For example, what is the probability of scoring a prime number when you throw a dice?

There are three possible acceptable outcomes (2, 3, 5) from a total of six possible outcomes (1, 2, 3, 4, 5, 6).

This could be written as: $P(\text{prime}) = \frac{3}{6} = \frac{1}{2}$

Probabilities can also be written as decimals or percentages: $P(\text{prime}) = \frac{1}{2}$ or 0.5 or 50%

If an outcome is completely unpredictable, or if each of the possible outcomes has an equal chance of happening, then you say that the outcome is **random**.

> **Revision tip**
>
> If you write a probability as a fraction, always write it down first then simplify it.

Probability that an event will not happen

A box contains a mixture of 15 white and 5 milk chocolates.

What is the probability of not picking, at random, a milk chocolate?

P(not milk chocolate) = $\frac{15}{20}$

Because the chocolates are either milk or white, this is the same as P(white chocolate).

In the same way P(not white chocolate) = P(milk chocolate) = $\frac{5}{20}$

You can see from these that:

P(white chocolate) + P(not white chocolate) = 1

… and also that:

P(not white chocolate) = 1 − P(white chocolate)

This gives the general rule that:

P(outcome not happening) = 1 − P(outcome does happen)

Given that the probability a netball team wins its next game is 0.6, what is the probability that it does not win its next game? 1 − 0.6 = 0.4

Probability in practice

Emma rolled a fair dice 300 times and recorded the number of times she scored each number. She recorded the results in a frequency table.

Number	1	2	3	4	5	6
Frequency	45	51	54	49	48	53

This table can be used to find the **experimental probability** or **relative frequency** for the event (throwing the dice).

The relative frequency of an outcome or event =
$\frac{\text{frequency of the outcome or event}}{\text{total number of trials}}$

Emma has an experimental probability for P(1) = $\frac{45}{300}$ = $\frac{3}{20}$

As the number of **trials**, or experiments, increases the experimental probability gets closer to the **theoretical probability**, P(1) = $\frac{1}{6}$

Emma's results are not too far away from results that theoretical probability would suggest:

P(1) = $\frac{1}{6}$ × 300 = 50 for each number.

> **Revision tip**
>
> If, after many trials, one of the numbers was very different to that expected, it might suggest that the dice was **biased**.

Combined events

Situations often occur in which two events happen together (**combined events**), so the probabilities need combining. For example, these two spinners are spun at the same time.

The products of the two outcomes are recorded in a **possibility diagram** (sometimes called a **sample space diagram**) like that shown on p145.

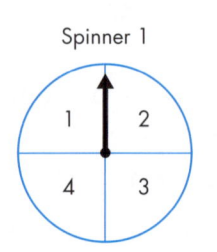

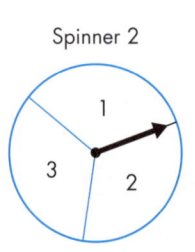

Spinner 1

Spinner 2

IGCSE Mathematics Revision Guide

		Spinner 2		
	Products (×)	1	2	3
Spinner 1	1	1	2	3
	2	2	4	6
	3	3	6	9
	4	4	8	12

There are 12 different outcomes, so any probability written as a fraction will have a denominator of 12

P(even number) = $\frac{8}{12} = \frac{2}{3}$, P(square number) = $\frac{4}{12} = \frac{1}{3}$, P(< 6) = $\frac{7}{12}$

The table is a very good way of representing the possible outcomes because you can see all possible combinations – it is an **exhaustive** list.

Tree diagrams

When Ben plays chess, the probability of him winning is $\frac{1}{2}$, of drawing is $\frac{1}{5}$ and of losing is $\frac{3}{10}$

These outcomes can be drawn as a **tree diagram**, which looks like the branches of a tree, with the probabilities written on the branches and the outcomes written at the end of the branches.

If he plays a second game, the tree can be extended to represent the second game with the probabilities written on the branches again.

Tree diagrams are used to find the probabilities of combined events. The probabilities are found by multiplying the probabilities of the individual outcomes, so that the probability of a win followed by a win is

P(win/win) = $\frac{1}{2} \times \frac{1}{2} = \frac{1}{4}$, and so on.

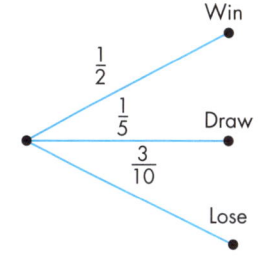

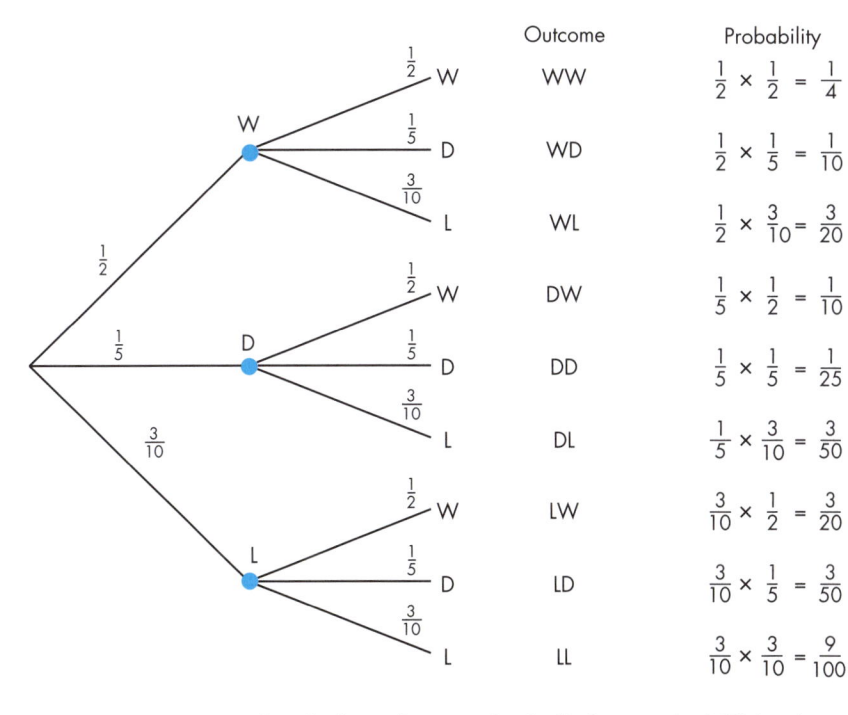

Outcome	Probability
WW	$\frac{1}{2} \times \frac{1}{2} = \frac{1}{4}$
WD	$\frac{1}{2} \times \frac{1}{5} = \frac{1}{10}$
WL	$\frac{1}{2} \times \frac{3}{10} = \frac{3}{20}$
DW	$\frac{1}{5} \times \frac{1}{2} = \frac{1}{10}$
DD	$\frac{1}{5} \times \frac{1}{5} = \frac{1}{25}$
DL	$\frac{1}{5} \times \frac{3}{10} = \frac{3}{50}$
LW	$\frac{3}{10} \times \frac{1}{2} = \frac{3}{20}$
LD	$\frac{3}{10} \times \frac{1}{5} = \frac{3}{50}$
LL	$\frac{3}{10} \times \frac{3}{10} = \frac{9}{100}$

> **Revision tip**
>
> To calculate probabilities of a combined event, multiply the separate probabilities along the branches. To calculate the probability of multiple combined events, add the probabilities in the list.

It is important to check that the total of all the probabilities is 1, $\frac{100}{100}$, since the list is exhaustive – no other combinations are possible.

The P(winning at least 1 game) = WW + WD + WL + DW + LW = $\frac{1}{4} + \frac{1}{10} + \frac{3}{20} + \frac{1}{10} + \frac{3}{20} = \frac{3}{4}$

Using Venn diagrams with probability

In the section 'Ordering and set notation' (page 21) you were introduced to Venn diagrams and the notation used with them. They can be useful in solving probability problems as well.

For example, in a survey, Sacha asked 100 people if they liked apples and bananas. The results are shown in the Venn diagram.

Make sure you understand these results:

$P(A) = \frac{52}{100}$

$P(Not\ A) = \frac{48}{100}$

$P(B) = \frac{65}{100}$

$P(A \cup B) = \frac{82}{100}$

$P(A \cap B) = \frac{35}{100}$

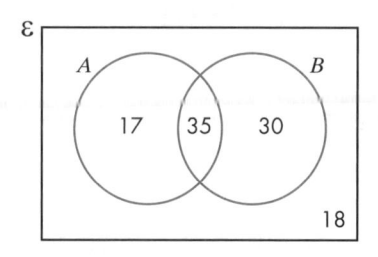

Extended

Conditional probability

Sometimes the probability of an event is dependent on the outcome of another event. If you take a card from a deck of cards and do not return it, the probabilities for the card you take next will change. This is **conditional probability**.

This can be shown with a probability tree or a Venn diagram.

Example 1: a physical education teacher has 5 volleyballs and 7 netballs in a sack. Two students each take a ball from the sack.

Work out the following probabilities:

P(both balls are volleyballs)

P(the balls are different)

Draw a tree to show this information.

Remember that the second student only has 11 balls to choose from and the number of volleyballs and netballs will also be different from the starting position.

P(both balls are volleyballs) = $\frac{20}{132}$

P(the balls are different) = $\frac{35}{132} + \frac{35}{132} = \frac{70}{132}$

Example 2: 50 students in Year 9 are choosing IGCSE options. 28 students choose Spanish and 25 choose French, with 5 students choosing neither language. One of the students in the Spanish class is chosen at random. What is the probability this student also studies French?

Draw a Venn diagram to show the information.

Be careful working out how many students have chosen both French and Spanish.

The probability calculation is the intersection divided by the set of the initial condition.

In this case this it is the intersection divided by the Spanish set.

$P(S) = \frac{P(F \cap S)}{P(S)} = \frac{8}{28}$

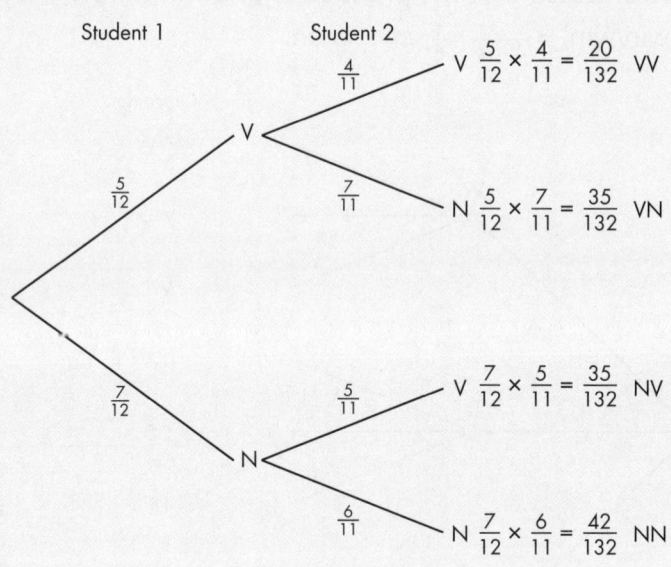

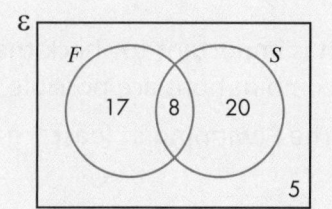

Quick test

1. Sven has 50 DVDs: 12 comedy, 4 music, 17 drama, 10 cartoons, 7 action. He takes a DVD from the shelf at random. What is the probability that the DVD is:
   a) drama
   b) not music
   c) comedy or action
   d) sport?

2. The probabilities of choosing, at random, a student from a particular tutor group in Year 11 are shown in this table.

Group	11AB	11MN	11PQ	11ST	11XY
Probability	0.2	0.21	x	0.23	x

   What the probability of choosing a student in 11PQ?

3. Alice takes a counter from a bag containing 20 counters. She notes its colour and replaces it in the bag. There are 12 white, 5 yellow and 3 blue counters. The table shows the results for different numbers of experiments.

Number of trials	10	50	100	250	500
Number of times a white counter is pulled out	4	31	48	130	295

   a) What is the theoretical probability of pulling out a white counter?
   b) Work out the experimental probability of pulling out a white counter for each of the number of trials.
   c) Which number of trials gives a probability closest to the theoretical probability? Why is this?

4. Spinner A has three equally spaced sectors. Two sectors are red and one sector is blue.
   Spinner B has four equally spaced sectors. One sector is red, two sectors are blue and one is green.
   Spinner A is turned and then spinner B is turned.
   Using either a possibility diagram or a tree diagram, work out these probabilities.
   a) P(spinning 2 blue sectors)
   b) P(spinning at least 1 red sector)
   c) P(not spinning a green sector)

5. The probability that Sabu will meet Tarit at the school bus stop on time is 0.8
   Draw a tree diagram to work out these probabilities.
   a) P(They meet on time on two consecutive days)
   b) P(They will both be on time for only one of two consecutive days)

6. a) Copy and complete the Venn diagram on the right, which contains the probabilities of events A and B occurring.
   b) Work out i) $P(B)$ ii) $P(A \cup B)$ iii) $P(A \cap B)$

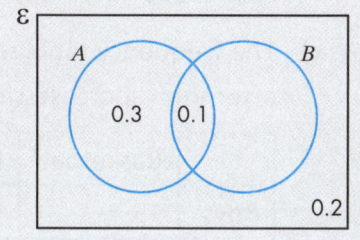

Extended

7. Ahmed arranges to play 2 games of chess with his friend Medhi. The probability of Ahmed winning the first game is 0.65 If Ahmed wins the first game, the probability of him winning the second game is 0.8 If Medhi wins the first game the probability Ahmed wins the second is 0.6
   Work out these probabilities:
   a) P(Medhi will win the first but lose the second game)
   b) P(Ahmed wins both games)
   c) P(Medhi will win only one game)

Exam-style practice questions

1 **a)** These are the makes of the first 60 cars to pass the college gate one Monday morning.

VW	Ford	Mercedes	Renault	Ford	Ford	Renault	Renault	Suzuki
Ford	Vauxhall	Peugeot	VW	Mercedes	Vauxhall	Ford	Peugeot	Vauxhall
Vauxhall	Peugeot	Peugeot	Vauxhall	Renault	VW	Ford	Renault	Mercedes
Suzuki	Renault	Ford	Ford	Ford	Suzuki	VW	VW	Ford
Renault	Suzuki	Peugeot	Renault	Vauxhall	Ford	Peugeot	Peugeot	Mercedes
Peugeot	Ford	VW	Suzuki	Ford	Peugeot	Mercedes	Peugeot	VW
VW	Vauxhall	Peugeot	Peugeot	Suzuki	Vauxhall			

Create a frequency table from this information.

What make of car is the mode? [2]

b) The pictogram shows the number of cakes sold by a baker during one a week.

Six cakes were sold on Thursday and 16 cakes were sold on Friday.

Copy and complete the pictogram to show the cake sales for Thursday and Friday. [2]

How many cakes were sold in the week, including those sold on Thursday and Friday? [2]

c) The frequency table shows the sports that students in a year group choose for their afternoon sports session.

	Basketball	Football	Hockey	Netball	Rounders
Boys	15	34	18	0	6
Girls	6	10	10	23	19

Draw a dual bar chart to represent this information. [5]

d) Use the data in the frequency table above to draw two pie charts that compare the sport choices made by boys and girls. [4]

148 IGCSE Mathematics Revision Guide

e) A breeder of exotic fish carried out a survey over eight years to investigate the average mass of koi carp at various ages. The data collected is given in the table.

Age of koi (years)	1	2	3	4	5	6	7	8
Average mass (g)	90	230	795	610	1050	1090	1280	1560

Plot the data on a scatter diagram. [4]

A mistake was made in calculating one of the average mass figures.

Which figure looks to be in error?

Draw a line of best fit and estimate the mass of a 9-year-old koi. [2]

f) The masses of 100 eggs are recorded in this frequency table.

Mass (g)	$50 < g \leq 52$	$52 < g \leq 54$	$54 < g \leq 56$	$56 < g \leq 58$	$58 < g \leq 60$	$60 < g \leq 62$	$62 < g \leq 64$
f	9	22	15	17	12	16	9

Draw a histogram for this data. [3]

g) Twenty nine people have joined a gym. As part of their registration application they were weighed. The results (in kg) are these:

62 73 92 64 78 70 82 92 61 81 74 83 86 86 88

67 75 82 93 62 66 71 96 87 86 74 85 79 78

Place and arrange these values into an ordered stem and leaf diagram and identify the median and mode weight of the new members. [3,2]

Extended

h) The table shows the number of DVDs (d) owned by 280 students.

DVDs (d)	$0 < d \leq 20$	$20 < d \leq 40$	$40 < d \leq 50$	$50 < d \leq 60$	$60 < d \leq 70$	$70 < d \leq 100$
F	16	34	68	89	54	24

Draw a histogram for this data. [4]

Estimate how many students had fewer than 25 DVDs. [2]

2 a) The scores for 12 students in a test were:

27 22 15 18 19 20 21 15 16 21 25 21

What are the mode, median and mean test scores? [1, 1, 2]

What is the range? [1]

b) The table shows the number of butterflies seen in a garden in a 15-minute period at noon over 50 consecutive days.

Butterflies (x)	10	11	12	13	14	15
Number of days (f)	4	7	12	11	9	7

What are the mode, median and mean numbers of butterflies? [1, 2, 3]

What is the range? [1]

Extended

c) This grouped frequency table shows the run score of a cricket team over a number of matches.

Run score (s)	0 ⩽ s < 50	50 ⩽ s < 100	100 ⩽ s < 150	150 ⩽ s < 200	200 ⩽ s < 250	250 ⩽ s < 300
Matches (f)	3	6	11	7	2	1

Work out an estimate of the mean number of runs (to 3 s.f.) and identify the class containing the median and the modal group. [3,2,1]

d) This grouped frequency table shows the number of hours that customers of an internet cafe spent logged onto the internet over a period of one week.

Hours	0 < t ⩽ 5	5 < t ⩽ 10	10 < t ⩽ 15	15 < t ⩽ 20	20 < t ⩽ 25	25 < t ⩽ 30	30 < t ⩽ 35
Frequency	4	9	15	8	7	4	1

Use the data in this table to draw a cumulative frequency diagram. [4]

Use the diagram to work out the median, the lower and upper quartiles and the interquartile range. [4]

e) Using your cumulative frequency diagram and results from question 2d) above, create a box and whisker diagram for the hours spent logged on to the internet. [3]

3 a) Gemma is asked to think of a number from 1 to 100 inclusive. Assuming her choice is random, what is the probability that her choice is:

i) 10 or less [1]   ii) higher than 80 [1]   iii) an even number [1]

iv) a square number [1]   v) 23 [1]   vi) a cube number? [1]

b) A letter is chosen at random from the word **MISSISSIPPI**. What is the probability that it is:

i) not I [1]   ii) S [1]   iii) not P [1]   iv) not a consonant? [1]

c) Mikhail makes an eight-sided spinner and spins it 200 times. Tanya makes a 10-sided spinner and spins it 250 times. Here are their results.

Spin score	1	2	3	4	5	6	7	8	9	10
Mikhail	24	18	22	26	24	36	27	23		
Tanya	28	24	25	29	27	26	25	24	20	22

Which of the two spinners appears to have the least bias? [1]

Justify your answer. [2]

d) A bag contains three yellow counters and two blue counters.

Omar takes one counter out of the bag at random. He notes the colour and puts it back in the bag, then takes another counter.

   i) Complete this probability space diagram. The first letter represents counter 1 and the second letter represents counter 2. [3]

		\multicolumn{5}{c}{Counter 1}
		Y	Y	Y	B	B
Counter 2	Y	YY				
	Y					
	Y			YY	BY	
	B				BB	
	B		YB			

   ii) Find P(2 yellow counters), P(2 blue counters), P(1 yellow and 1 blue counter), P(blue and then a yellow). [4]

e) A game at a fairground involves throwing tennis balls at a target. Players get two throws. If a target is hit they get a prize. The probability of a player hitting a target with the first throw is 75%. If a target is hit with the first throw, the player has a probability of 90% of hitting a second target. If a player does not hit a target with their first throw, they have a 40% probability of hitting a target with the second throw.

Draw a tree diagram to represent this information. [3]

   i) What is the probability of a player hitting two targets? [1]

   ii) What is the probability of a player hitting only one target? [1]

   iii) 3000 play the game in one day. How many players are likely to leave the game without having hit a target? [2]

f) There are 120 skiers waiting at a ski lift. 60 have a season pass, 50 are snowboarders and 15 of these snowboarders have a season pass.

   i) Draw a Venn diagram to represent this information. [3]

   ii) Work out the probability that a person chosen at random from the 120 skiers is a season pass holder but they are not a snowboarder. [1]

Extended

g) There are 20 people queuing at a lift at a ski resort. 14 people are skiers and the other 6 are snowboarders. They get into the lift one at a time.

What is the probability that:

   i) the first two people who get into the lift are a skier and a snowboarder? [1]

   ii) at least one snowboarder gets into the lift? [1]

Answers

Number (p. 9)

Quick test
1. a) 1 and 9   b) 1 and 8   c) 3
2. 8
3. $40 = 2^3 \times 5$, $56 = 2^3 \times 7$, LCM $= 2^3 \times 5 \times 7 = 280$
4. For example $\sqrt{4} = 2$ and 2 is a rational number
5. a) 0.0625   b) 2.5   c) 1.6   d) 0.3
   e) $\frac{3}{11}$

Fractions and percentages 1 (p. 12)

Quick test
1. $\frac{4}{7} = \frac{12}{21} = \frac{16}{28} = \frac{24}{42}$
2.

Decimal	Fraction	Percentage
0.875	$\frac{7}{8}$	87.5%
0.3125	$\frac{5}{16}$	31.25%
0.15	$\frac{3}{20}$	15%

3. a) 263.2 km   b) $2975
4. 6%

Fractions and percentages 2 (p. 15)

Quick test
1. $\frac{8}{15} = 0.5\dot{3}$, $\frac{4}{7} = 0.\dot{5}7142\dot{8}$, $\frac{5}{12} = 0.41\dot{6}$
2. $\frac{13}{55}$
3. $500
4. A gives €1800 simple interest and B gives €1968.74 compound interest, so Plan B is the better one.
5. $11 300

The four rules (p. 17)

Quick test
1. a) $(5 + 2^2) \div 3 = 3$
   b) $3^3 \times (6 - 2) = 108$
2. $7.50
3. a) $\frac{629}{40}$ b) $1\frac{1}{40}$ c) $10\frac{3}{20}$ d) $4\frac{1}{2}$
4. $225

Directed numbers (p. 18)

Quick test
1. a) +1   b) +6   c) +9   d) +9

2. $6 \times (4 \div -2)$ [−12], $30 \div (-2 \times -3)$ [5], $-4 + 10 - -7 - 5$ [8], $5 \times (30 \div (5 - -7))$ [15]

Squares and cubes (p. 20)

Quick test
1. a) 361   b) 23   c) 456   d) 274.625
2. $\sqrt[3]{32768} = 32$, all the others evaluate to 64
3. a) 5^6 by 7849   b) $\sqrt[8]{6561}$ by 1
4. 558
5. 551 400

Ordering and set notation (p. 22)

Quick test
1. a) $\sqrt[3]{75} > 2.05^2$ b) 0.2 litres < 200 cl
   c) $\frac{3}{8}$ of 200 km $> \frac{2}{5}$ of 185 km
2. −6, −5, −4, −3, −2, −1, 0, 1, 2
3. a) i) {15}
      ii) {3, 5, 6, 9, 10, 12, 15, 18, 20}
   b) $A \cap B = \{15\}$, $n = 1$
   c) 0. This is an empty set – there is no intersection between sets B and C.
   d) i) {2, 8, 13, 17}
      ii) {2, 5, 8, 10, 13, 15, 17, 20}
   e) $n(A \cap B)'$ is everywhere outside intersection of A and B.

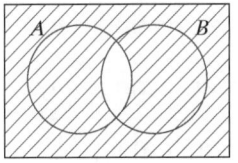

   $n(A \cap B)' = \{2, 3, 5, 6, 8, 9, 10, 12, 13, 17, 18, 20\}$ $n = 12$
   f) (Venn diagram)

Ratio, proportion and rate 1 (p. 24)

Quick test
1. a) 2 : 5   b) 10 : 3
   c) 625 : 1   d) 1 : 27
2. $\frac{3}{8}$
3. 90
4. 1.2 km

Extended
5. 160 mm
6. 10:25 am

Ratio, proportion and rate 2 (p. 26)

Quick test
1. $50.40, 8
2. 11
3. 120
4. 6.75 cm
5. a) 2.5 g/cm³   b) 2.2 g/cm³
6. 319 000 Pa

Estimation and limits of accuracy 1 (p. 28)

Quick test
1. a) 23.76   b) 21.3   c) 34.90   d) 0.056
2. $7250 \leq$ attendance < 7255
3. a) 23.8   b) 20   c) 34.90   d) 0.06

Estimation and limits of accuracy 2 (p. 30)

Quick test
1. No, the fastest that Rahim could be is 13.555 s, which is not as fast as Joshua, which could be 13.550 s.
2. 91.75 m and 50.25 m
3. 5247.9 kg
4. 8 mm
5. Upper bound 101 trees; lower bound 95 trees. Using just the upper bound measurement, then yes the woodland could be classed as a forest.

Standard form (p. 32)

Quick test
1. a) 3.17×10^2 b) 3.17×10^4
   c) 3.17×10^{-1} d) 3.17×10^{-4}
   e) 3.17×10^0
2. a) 487 000   b) 487
   c) 0.0487   d) 0.000 004 87
3. a) 3.8×10^6 b) 2.128^5
   c) 2.45×10^{-6} d) 8×10^0
   e) 6.24×10^{-7} f) 2.05×10^2
4. Approximately 449 days
5. 90 sheets of gold leaf ≈ 1 sheet of printer paper

152   IGCSE Mathematics Revision Guide

Applying number and using calculators (p. 36)

Quick test
1. a) centimetres b) tonnes
   c) millilitres (or centilitres)
2. a) 0.23 m b) 1050 ml
   c) 0.205 kg d) 350 000 mm³
3. 10:30
4. 37 hours, 15 minutes
5. 2750 krone, 45 krone
6. a) 1796.978544 b) 2.216304326

Exam-style practice questions (p. 37–40)

1. a) $2^2 \times 7^2 \times 11$
   b) $2156 = 2^2 \times 7^2 \times 11$, $385 = 5 \times 7 \times 11$, HCF $= 7 \times 11 = 77$
   c) $3^2 = 9$ and $4^2 = 16$, so any of $\sqrt{10}, \sqrt{11}, \sqrt{12}, \sqrt{13}, \sqrt{14}, \sqrt{15}$
   d) When you divide by a fraction you invert the fraction and then multiply by it. To form a reciprocal of a fraction, you invert it, so what Yumi says is a description of what you do to carry out the division.
2. a) $\frac{19}{4} = \frac{627}{132}, \frac{25}{6} = \frac{550}{132}, \frac{45}{11} = \frac{540}{132}$. So, $\frac{19}{4}$ is the largest.
   b) If $x = 0.02\dot{5}$, $1000x = 25.\dot{2}\dot{5}$ and $10x = 0.\dot{2}\dot{5}$, $\frac{25}{990} = \frac{5}{198}$
   c) $475 \times 0.012 = \$5700$
   d) $\frac{10899 - 10500}{10500} \times 100 = 3.8\%$
   e) 2h 30m = 150m, 0.82 × 150 = 123m = 2h 3m
   f) $25000 \times 1.0125 = \$26536.43$
   g) $\frac{1766.4}{138} \times 100 = \$1280$
3. a) $\frac{384}{65} \approx 6$ coaches, $\frac{(6 \times 350) + 75}{384} \approx \$5.67$
   b) $4\frac{2}{7} + 5\frac{1}{6} + 3\frac{2}{3} - 12\frac{1}{2} = \frac{13}{21}$ kg
   c) $11 \div 1\frac{7}{8} = 5\frac{13}{15} \approx 5$ units, $1\frac{7}{8} \times \frac{13}{15} = 1\frac{5}{8}$
4. a) $24 - (-21) = 45°C$
   b) $-5 + 18 - 3 + 8 = 18$
5. a) $\sqrt{280}, \sqrt{300}, (-16.8)^2, 17^2$
   b) $\sqrt[3]{-400} = -7.37$, -7 and -8
   c) $900 > 1800 \times 0.925^x$, $0.5 > 0.925^x$, by trial and improvement $n = 9$
6. a) 31, ~~33~~, 35, 37, ~~39~~
   b) i) {12} ii) {3, 4, 6, 8, 9, 12}
   c) 1
   d) i) $(A \cup B)' = \{1, 2, 5, 7, 10, 11\}$
      ii) $A \cap B' = \{3, 6, 9\}$
   e) 83, 89, 97 → {83, 89}, {83, 97}, {89, 97}, {83}, {89}, {97}, {} → 7
7. a) $8 \div 5 = 1.6$, $1.6 \times 3 = 4.8$, $8 + 4.8 = 12.8$ litres
   b) $\frac{268800}{12} \times 11 = 246400$
   c) $1.5 + 40 + 10 = 51.5$ km, $17 + 55 + 31 + 103$ minutes, $\frac{51.5}{103} \times 60 = 30$ km/h
   d) i) $150 \times 632 \div 1000 = 94.8$ tonnes
      ii) $632 \div 94.8 = 6\frac{2}{3}$ km/tonne
      iii) No. $220 \times 6\frac{2}{3} \approx 1467$ km
   e) $210 \times 210 = 44100$ cm², $60 \times 20 \times 100 = 1200000$ ml, $1200000 \div 44100 \approx 27.2$ cm
      $210 \times 210 \times 100 = 4410000$ cm³ (ml), $4410000 \div 1000 \div 20 = 220.5$ minutes
   f) Pete: $\frac{16}{25} = 64$ p/kg, Kirsty: $\frac{12.6}{20} = 63$ p/kg, so Kirsty's is the best buy.
   g) $500000.5 = 25000$ p, $25000 \div 1.25 = 20000$ times
   h) $2.5 \times 2.5 \times 10 = 6250$ cm³, $4700 \div 6250 = 0.752$, so Kai is correct.
   i) The face with the largest area, 1.5 m × 1 m
8. a) 530, 527.82, 527.8, 528
   b) 14.95 kg ⩽ sand < 15.05 kg
   c) 462.5 ml ⩽ ladle < 487.5 ml, 24950 ml ⩽ pot < 25050 ml
      Min = $\frac{24950}{487.5} = 51.179 \approx 51$ servings.
      Max = $\frac{25050}{462.5} = 54.16 \approx 54$ servings
9. a) 2.54×10^7, 2.54×10^{-4}
   b) Mars, 1.1974×10^8 (119 740 000) km
10. a) $(1500 \times 1.37 + 750) \div 1000 = 2.805$ kg
    b) $(2500 \times 24850) - (315 \times 25650) = 54045250$ dong
    c) 1.849790551

Algebraic representation and formulae (p. 43)

Quick test
1. a) $22.50 b) $7.5x c) $xy
2. $k - x$ years
3. a) $2b$ b) $2b + 5$ c) $2b + 3y + 5$
4. a) 13 b) 12.21
5. a) $90
   b) $H = \frac{C - 1.25D - 25}{4}$, 25 hours
   c) $D = \frac{C - 4H - 25}{1.25}$, 50 km

Algebraic manipulation 1 (p. 46)

Quick test
1. a) $4v$ b) $8pw$ c) $20q^2$
   d) $4fg^2h$ e) $-6d^2e$
2. a) $9p + 2r$ b) $4cd - 3tv$
   c) $7w^2$ d) $g^3 - a^2$
3. a) $4h + 8$ b) $6v - 4tv$
   c) $-2q - 2qr - 5v$ d) $8gh + 9g^2$
4. a) $4(4p - 3q)$ b) $3s(4 + 3t)$
   c) $2g(10gh - 7f)$ d) $3cd^2(3 - 5cde)$
   e) $8mn(2n^2 + 1)$
5. a) $x^2 + 4x - 12$ b) $6w^2 - 8w - 8$
   c) $p^2 + 6p + 9$ d) $12 + 14m - 6m^2$
   e) $q^2 - 4$
6. $x^3 - 6x^2 + 12x - 8$

Algebraic manipulation 2 (p. 48)

Quick test
1. a) $(m - 6)(m + 4)$ b) $(3 - y)(2 + y)$
   c) $(v - 1)(v - 3)$ d) $(w + 4)(w + 4)$
   e) $(p + 3)(p - 6)$
2. a) $(p + 7)(p - 7)$ b) $(q + 11)(q - 11)$
   c) $(w - 1)(w + 1)$ d) $(15 - d)(15 + d)$
   e) $(e - f)(e + f)$
3. a) $\frac{13x}{15}$ b) $\frac{19 - 3x}{10}$ c) $\frac{1}{6}$
   d) $\frac{8}{25y}$ e) $\frac{x + 2}{2x - 3}$

Solutions of equations and inequalities 1 (p. 51)

Quick test
1. a) $w = 4$ b) $t = 2$ c) $v = \frac{1}{2}$
   d) $d = -2$
2. $m = 20$ minutes, so an 18 kg turkey would require 7 hours 30 minutes cooking time.
3. a) $w > -1$ b) $v \leqslant 2$ c) $t \geqslant -1$
   d) $x < 4$
4. a) number line from 2 to 5, open circle at 3
   b) number line from 0 to 3, closed circle at 2

Simultaneous equations (p. 53)

Quick test
1. a) $x = \frac{1}{2}$ and $y = \frac{3}{4}$
   b) $x = -3$ and $y = 2.5$
   c) $x = 2$ and $y = 3$
   d) $x = 4$ and $y = -\frac{1}{2}$
2. $43.50
3. $x = 1, y = 2$, and $x = 2, y = 1$

Solutions of equations and inequalities 2 (p. 55)

Quick test
1. a) $x = -4$ or -7 b) $z = -3$ or $+5$
   c) $s = 3$ or -2 d) $q = \pm\frac{4}{3}$
2. a) $x = 0.39$ or -10.39
   b) $x = 7.16$ or 0.84
   c) $x = 6.16$ or -0.16 d) $x = 0$ or -3.6
3. a) $x = 0$ or -10 b) $x = -5 \pm \sqrt{29}$
   c) $x = 2 \pm \sqrt{10}$
   d) $x = 7.5 \pm \sqrt{54.25}$

Answers 153

4. a) $x = 0$ or -12
    b) $x = 0.71$ or -12.71
    c) $x = 4.68$ or 0.32
    d) $x = 12.56$ or -0.56

Graphs in practical situations (p. 58)

Quick test
1. a) 9 kg
    b) Yes. Although 20 kg is not on the graph, you could assume that a 20 kg turkey takes twice as long as a 10 kg turkey, but remembering to subtract $1\frac{1}{2}$ hours from the total time. This is added to each cooking time and you have one 20 kg turkey and not two 10 kg ones.
2. a) 53 km/h   b) Approx. 28 km
    c) 16 km/h
3. a) 0.64 m/s^2 b) 19.5 km/h
    c) 0.25 m/s^2 d) 125 m
4. 52

Straight-line graphs (p. 62)

Quick test
1. AB has gradient $m = 0.5$, CD: $m = -2$
2. $y = x - 1$
3. $y = 2 - 2x$
4. AB: $y = 0.5x + 3.5$, CD: $y = 8 - 2x$
5. AB: midpoint $(1, 4)$,
    CD: midpoint $(0.5, 6.5)$
6. $y = 0.5x - 3$
7. $y = -3 - 2x$
8. $y = -4 - 2x$
9. $y = 0.5x - 1$
10. Yes. All four lines are either parallel or perpendicular to each other. A rectangle has two sets of parallel lines with four angles of 90°.
11. $y = 11 - 2x$

Graphs of functions (p. 67)

Quick test
1. a) $(2, -5)$ b) -0.8 and 4.8
    c) -4.75 d) -0.24 and 4.24

2. $x = 3$

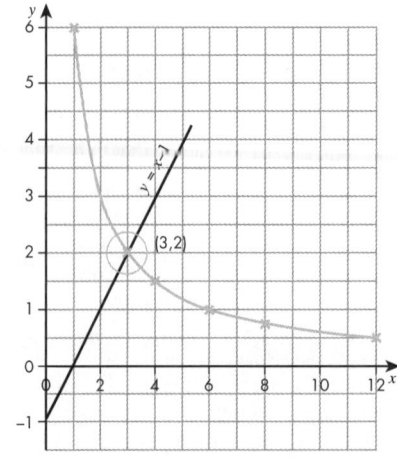

3. a)

x	15	16	17	18	19	20
y	880	1100	1300	1500	1800	2100

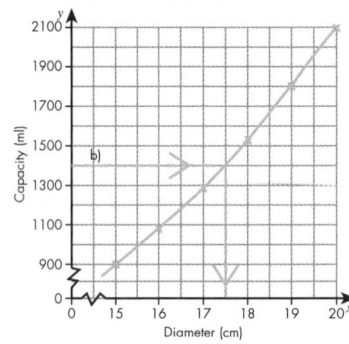

    b) Diameter 17.5 cm

4. a)

x	0	1	2	3	4	5
y	20000	14000	9800	6900	4800	3400

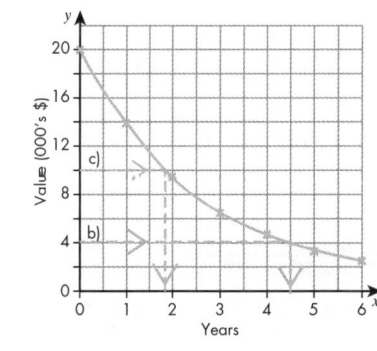

    b) 2 years   c) 4.5 years
5. a) ≈ 1.3 b) ≈ 0.225
6. Turning point $(-2, -11)$ and the roots are $x = -2 \pm \sqrt{11}$

Number sequences (p. 71)

Quick test
1. a) 4, 5, 6, 7, 8
    b) $-1, 1, 3, 5, 7$
    c) 5, 14, 29, 50, 77
    d) 23, 20, 15, 8, -1
2. a) 46, 53, 60, $7n + 4$
    b) 23, 27, 31, $4n - 1$
    c) 22, 13, 4, $76 - 9n$
    d) $-11, -17, -23, 25 - 6n$

3. a)

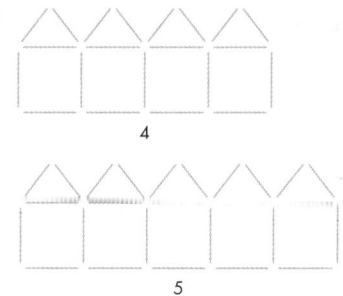

    b) $5n + 1$ c) 176   d) 10
4. a) 768   b) 607.5
5. a) 1.5×6^n b) 2×3^n
6. a) 30   b) 435

Indices (p. 73)

Quick test
1. a) 625   b) $0.0625 \left(\frac{1}{16}\right)$
    c) 6   d) 1
2. a) 3^7 b) 5^2
3. a) 9   b) 125   c) $\frac{1}{64}$ d) $\frac{4}{25}$
4. a) $\frac{3}{a^2}$ b) $\frac{6}{a^2}$
    c) $\frac{1}{2a^2}$ d) $\frac{4}{5b^3}$
5. a) $5v^{-2}$ b) $3w^{-6}$ c) $2t^{-4}$ d) za^{-2}
6. a) $6m^3$ b) $4n^6$
7. a) $e^{\frac{5}{6}}$ b) $f^{\frac{7}{6}}$ c) $g^{-\frac{11}{6}}$ d) $h^{\frac{8}{3}}$
8. a) $4q^{\frac{1}{3}}$ b) $s^{\frac{4}{3}}$ c) $d^{\frac{9}{4}}$ d) $\frac{h^{\frac{9}{4}}}{8}$
    e) $q^{\frac{7}{12}}$

Variation (p. 75)

Quick test
1. a) 54   b) 4
2. a) 100   b) 4
3. a) 0.5   b) 8

Linear programming (p. 78)

Quick test
1. a)

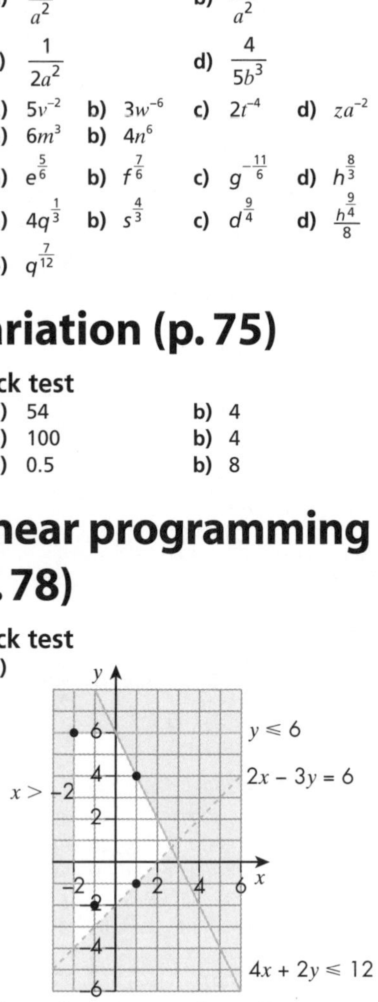

    b) $y = 6$ c) $x = 2$
    d) $(-1, -2)$
2. a) $x + y \leq 30$
    $1.5x + 3y \geq 60$
    $y \leq 18$
    $x \leq 14$

b) [graph showing $x + y \leq 30$, $x \leq 14$, $y \leq 18$, $1.5x + 3y \geq 60$]

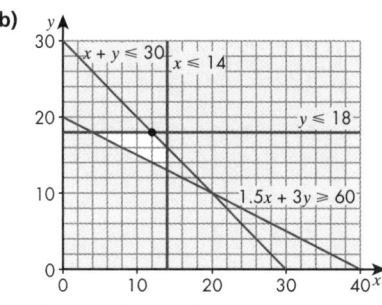

c) 12 of loaf $x \to 18$
   20 of loaf $y \to 60$
   = $78

Functions (p. 80)

Quick test
1. a) 4 b) 68 c) 14 d) –3.5
2. a) i) $f^{(-1)}(x) = \dfrac{x+7}{2}$ ii) $f^{(-1)}(x) = 3(x-4)$
   iii) $f^{(-1)}(x) = \dfrac{6-x}{x}$ iv) $f^{(-1)}(x) = \dfrac{5x+3}{2}$
   b) i) 5.5 ii) 0 iii) 0.5 iv) 11.5
3. a) 14 b) 13 c) 6
4. a) 17 b) 5 c) –5.5 d) 3
5. $ts(x) = \sqrt{5x-3}$

Exam-style practice questions (p. 81–86)

1. a) $c = 15t + 20j + 1.5s$
   b) i) $8.95
      ii) $13.95 = 1.25m + 2.7$, $11.25 = 1.25m$, $m = 9\,\text{kg}$
   c) i) $k = \dfrac{c - s^2}{1.25}$ ii) $s = \sqrt{c - 1.25k}$
   d) i) $v = \dfrac{1}{3}\pi r^2 h$, $3v = \pi r^2 h$, $h = \dfrac{3v}{\pi r^2}$
      ii) $v = \dfrac{1}{3}\pi r^2 h$, $3v = \pi r^2 h$, $r^2 = \dfrac{3v}{\pi h}$, $r = \sqrt{\dfrac{3v}{\pi h}}$

2. a) i) $12pq$ ii) $5m^4$
   b) $2x + 38$
   c) i) $6v + 12w$ ii) $8mp^2 - 12m^2$
   d) i) $10m - 6$
      ii) $3p(2q + p^2) - 5p^2(q - 2p) \to 6pq + 3p^3 - 5p^2q + 10p^3 \to 6pq + 13p^3 - 5p^2q$
   e) i) $3(2v - 5w)$
      ii) $3b^2c(2a + 5bc^2)$
   f) i) $d^2 - 3d - 10$ ii) $30 - e - e^2$
   g) i) $8g^2 + 14gh - 15h^2$
      ii) $17f - 12f^2 - 6$
   h) i) $25m^2 - 4n^2$ ii) $p^2q^2 - s^2t^2$
   i) i) $36a^2 - 60a + 25$
      ii) $b^2 - 6b + 8$

Extended
   j) i) $(c + 2)(c - 4)$ ii) $(f - 3)(f - 2)$
   k) Recognise the difference of 2 squares in these expressions:
      i) $(b + 7)(b - 7)$
      ii) $(5e + 10g)(5e - 10g)$
   l) i) $3(2x - 1)(x + 1)$
      ii) $(5x + 2)(x - 3)$
   m) i) $\dfrac{2}{x+3} + \dfrac{1}{2x-3}$, $\dfrac{2(2x-3) + (x+3)}{(x+3)(2x-3)}$, $\dfrac{4x - 6 + x + 3}{(x+3)(2x-3)}$, $\dfrac{5x-3}{(x+3)(2x-3)}$
      ii) $\dfrac{x^2 - 2x - 3}{2x^2 - 2x - 12}$, $\dfrac{(x-3)(x+1)}{(2x+4)(x-3)}$, $\dfrac{x+1}{2x+4}$
   n) $(x+3)(2x-1)(x+2)$, $(x+3)(2x^2 + 3x - 2)$, $2x^3 + 3x^2 - 2x + 6x^2 + 9x - 6$, $2x^3 + 9x^2 + 7x - 6$

3. a) i) $4d + 7 = 27$, $4d = 20$, $d = 5$
      ii) $20 - 7c = 6$, $14 = 7c$, $c = 2$
   b) i) $2m + 8 = 14$, $2m = 6$, $m = 3$
      ii) $9p + 30 = 12$, $9p = -18$, $p = -2$
   c) i) $q = 5$ ii) $s = -3$
   d) Starting position: $\dfrac{92}{2} \pm 5$, 46 ± 5, $A = 51$ and $B = 41$. Departing position: $A = 51 - 7 = 44$, $B = 41 + 7 = 48$

Extended
   e) i) $x^2 + 3x - 28 = 0$, $(x - 4)(x + 7) = 0$, $x - 4 = 0$ or $x + 7 = 0$, $x = 4$ or -7
      ii) $x^2 - 7x = -10$, $x^2 - 7x + 10 = 0$, $(x - 2)(x - 5) = 0$, $x = 2$ or 5
   f) i) $6x^2 - 21x = 0$, $3x(2x - 7) = 0$, $3x = 0$ or $2x - 7 = 0$, $x = 0$ or 3.5
      ii) $9x^2 + 7x + 4 = 6 + 2x - 3x^2$, $12x^2 + 5x - 2 = 0$, $(4x - 1)(3x + 2) = 0$, $x = \dfrac{1}{4}$ or $-\dfrac{2}{3}$
   g) i) $x(x + 6) = 21.5$, $x^2 + 6x - 21.5 = 0$ Use the quadratic formula to find $x = 2.5$ or -8.5 Length x cannot be negative so the wall is 2.5 m high and 8.5 m wide.
      ii) No. Each tile has side length $\sqrt{225} = 15\,\text{cm}$. $16\dfrac{2}{3}$ tiles are needed from floor to ceiling and $56\dfrac{2}{3}$ tiles are needed across the width, so tiles will need to be cut.
   h) i) $(x + 4.5)^2 - 26.25$
      ii) $3x^2 - 12x - 30 = 0$, $x^2 - 4x - 10 = 0$, $(x - 2)^2 - 14 = 0$, $(x - 2)^2 = 14$, $x - 2 = \pm\sqrt{14}$, $x = 2 \pm \sqrt{14}$
   i) Add the equations together to get: $5x = 20$, $x = 4$. Substitute $x = 4$ into 1 of the equations to find $y = -1$
   j) Double the second equation: to get the same coefficients for x: $2x + 4y = -4$. Subtract this new equation from the first to get: $4y = 2$, $y = 0.5$ Substitute $y = 0.5$ into one of the equations to find $x = -3$
   k) $x = \dfrac{3}{4}$, $y = -2$
   l) $S + R = 16$ and $S + 4 = 2(R + 4)$, rearrange this second to get $S - 2R = 4$. Use $S + R = 16$ and $S - 2R = 4$. Solve to find that Ruby is 4 years old and Sally is 12
   m) i) $x = 6$ ii) $x \leq 21$
   n) $x^2 + (3x - 10)^2 = 20$, $x^2 + 9x^2 - 60x + 100 = 20$, $10x^2 - 60x + 80 = 0$, $x^2 - 6x + 8 = 0$, $(x - 2)(x - 4) = 0$, $x = 2$ or 4 Substitute both back into one of the equations to find that when $x = 2$, $y = -4$ and when $x = 4$, $y = 2$
   o) $3(72 - 4x) < 84$, $72 - 4x < 28$, $44 < 4x$, $11 < x$
      $3(72 - 4x) > 0$, $72 - 4x > 0$, $72 > 4x$, $18 > x$ Combine the answers: $11 < x < 18$

4. a) i) [graph of Litres vs Gallons]
      ii) 15 Gallons = 67.5 Litres
   b) [graph of Distance (Km) vs Time (hours), Average speed ≈ 95 km/h]
   c) Distance is the area under the graph. Split the shape into manageable triangles, rectangles and trapezia, work out the area of each and add them together to get 400. The distance travelled is 400 km.

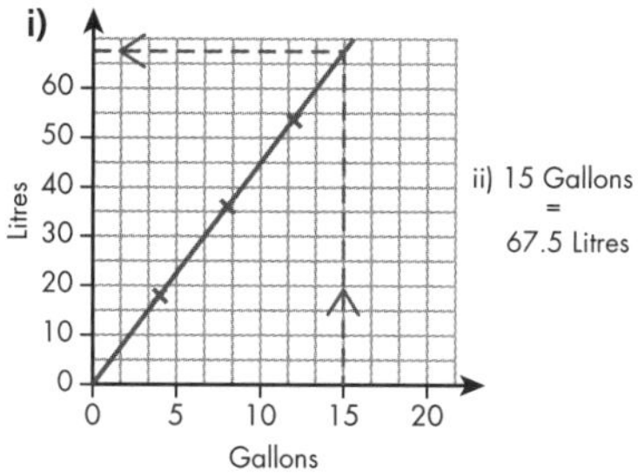

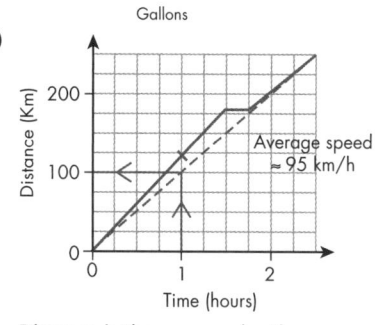

5. a) [graph showing $y = 2x - 2$ and $y = \dfrac{1}{2}x + 4$ intersecting at (4, 6)]
   b) $10 = 4m - 2$, $12 = 4m$, $m = 3$
   c) The gradient of the first line: $7 = 2m + 1$, $m = 3$. Because the lines are parallel the gradient of the second line is also 3, so $6 = 3(3) + c$, $c = -3$. The equation is: $y = 3x - 3$
   d) i) For the line segment AB: gradient $= \dfrac{6}{2} = 3$ Using (4, 8) in $y = mx + c$, $8 = 3(4) + c$, $c = -4$ AB has the equation, $y = 3x - 4$
      For the line segment AC: gradient $= \dfrac{-4}{4} = -1$ Using (4, 8) in $y = mx + c$, $8 = -1(4) + c$, $c = 12$ AC has the equation, $y = 12 - x$
      ii) (5, 3)
   e) $y = 12 - 2x$

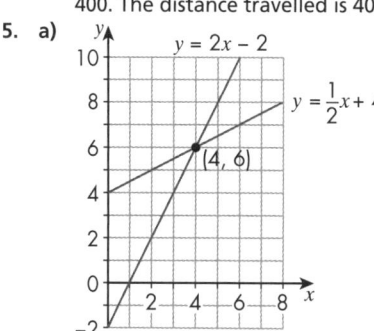

6. a) i)

x	–4	–3	–2	–1	0	1	2
y	5	0	–3	–4	–3	0	5

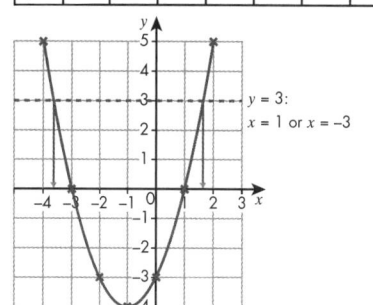

[parabola graph with horizontal line $y = 3$: $x = 1$ or $x = -3$]

ii) (−1, −4)

b)

x	0.5	1	2	4	8	16
y	8	4	2	1	0.5	0.25

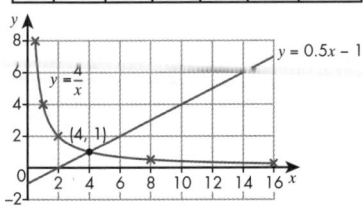

c)

x	−3	−2	−1.5	−1	−0.5	0	1
y	−5	1	$\frac{11}{8}$	1	$\frac{6}{5}$	1	7

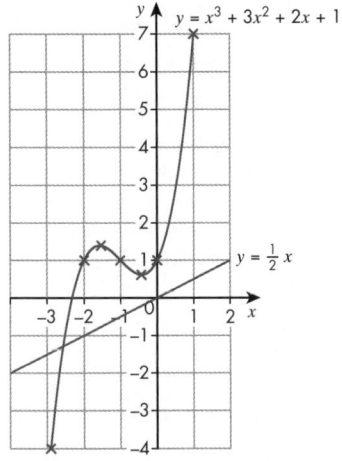

Stan is wrong. The line $y = 0.5x$ only intersects the cubic function at one point ≈ (−2.5, −1.3)

d)

Year	2016	2018	2020	2022	2024	2026
population	25000	20000	16000	13000	11000	8700

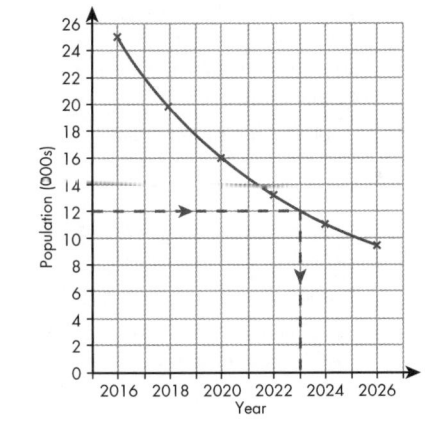

Population estimated to drop below 12 000 in 2023.

e) Gradient
at (−2,1) = 3 ÷ 1.5 = 2
at (0,1) = 3 ÷ 1.5 = 2

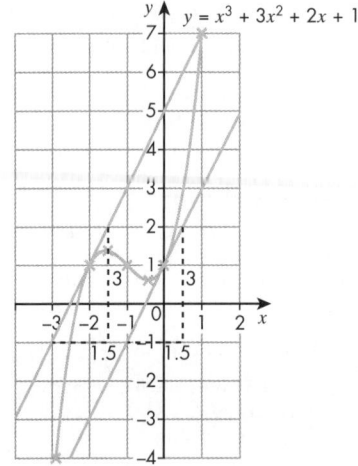

7. a) i) Add 4: 19, 23
   ii) Divide by 5: 0.32, 0.064
   b) i) 5, 7, 9, 11
   ii) −2, 1, 6, 13
   iii) 0.2, 1.6, 5.4, 12.8
   c) i) Add 4 to each term (4n) and the zero term is −2. The nth term is $4n − 2$
   ii) Compare known sequences against the question sequence. Notice that the terms in the sequence are all 3 greater than the terms in the sequence of square numbers so the nth term is $n^2 + 3$
   iii) The terms in the question sequence are all double the terms in the sequence of cube numbers so the nth term is $2n^3$
   d) i) 21, 31
   ii) 7th pattern has 43. Formula gives correct result.
   iii) 9901
   e) i) This is a quadratic sequence. The first differences are: 9, 13, 17, 21. The second difference is 4. This gives values of $a = 2$, $b = 3$ and $c = −4$ The nth term is: $2n^2 + 3n − 4$
   ii) $2(100)^2 + 3(100) − 4 = 20296$
   iii) $5351 = 2n^2 + 3n − 4$, $0 = 2n^2 + 3n − 5355$ Use the quadratic formula to find that $n = −52.5$ and 51 For a term to be in the sequence n should be a positive integer, so 5351 is the 51st term of the sequence.
   3361 is not in the sequence ($n = −41.78$ or 40.28)
   f) i) The sequence is the 3 × cubic numbers, so the nth term is: $3n^3$
   ii) $x_{10} = 3(10)^3 = 3000$, $x_{25} = 3(25)^3 = 46875$

8. a) i) 1024   ii) 53.1441
      iii) 1   iv) 7.2
   b) i) $\frac{1}{7^3}$ ii) $\frac{6}{b^2}$ iii) $5c^{-4}$
   c) i) $18f^5$
      ii) $9v^6$
      iii) $3mn^{-1}$ or $\frac{3m}{n}$
   d) i) 11   ii) 17
      iii) $\frac{11}{33}$ iv) $\frac{1}{7}$

   e) i) $w^{-\frac{3}{4}}$ ii) $x^{\frac{3}{5}}$
   f) i) 27   ii) 64
      iii) $y^{\frac{3}{4}} \times y^{-\frac{1}{2}} \div y^{\frac{1}{3}} = y^{\frac{3}{4}-\frac{1}{2}-\frac{1}{3}} = y^{-\frac{1}{12}}$
      iv) $\sqrt[3]{27} \times \sqrt[3]{z^2} = 3z^{\frac{2}{3}} \times$
      v) $x^{-\frac{1}{3}} = 2x^{-1}$, $x^{-\frac{1}{3}} = \frac{2}{x}$, $x \times x^{-\frac{1}{3}} = 2$,
      $x^{\frac{2}{3}} = 2$, $x = 2^{\frac{3}{2}}$, $x = \sqrt{8}$

9. a) i) $\frac{3.5}{24} \times 60 = 8.4$ kg
   ii) $\frac{25}{3.5} \times 5 = 35.7$ therefore 35 complete statues
   b) i) $36 = k \times 2^3$, $k = 4.5$, $4.5 \times 4^3 = 288$ g
   ii) $\sqrt[3]{121.5 \div 4.5} = 3$ cm
   c) i) $200 = \frac{k}{1250}$, $k = 250000$, $s = \frac{250000}{625}$, $s = 400$
   ii) $I = \frac{250000}{1000}$, $I = 250$

10. a) [graph: $y = 2 − 0.4x$]
    b) [graph with $y = 2x + 3$, $y = 0.5x + 1$, $2y + 3x = 9$]
    c) i) $x + y \leq 10$, $8x + 15y \leq 120$, $y > 4$, $x > 2$
    ii) [graph]
    iii) (3 small, 5 large) (3 small, 6 large) (4 small, 5 large) (5 small, 5 large)

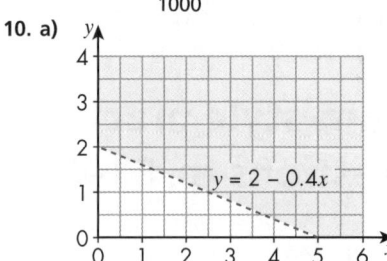

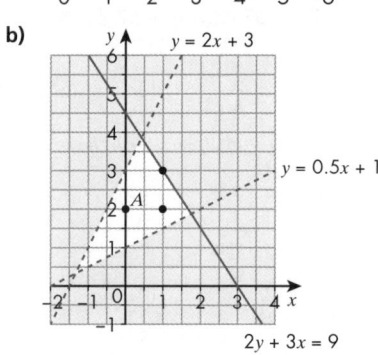

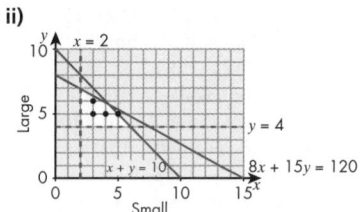

11. a) i) 47, 0
    ii) $29 = 2x^2 − 3$, $32 = 2x^2$, $10 = x^2$, $x = \pm\sqrt{16}$, $x = \pm 4$
    iii) $2x^2 − 3 = 6x − 3$, $2x^2 − 6x = 0$, $x(2x − 6) = 0$, $2x − 6 = 0$, $x = 3$ or $x = 0$
    b) i) $y = \frac{3}{2x-1}$, $2x − 1 = \frac{3}{y}$, $2x = \frac{3}{y} + 1$, $x = \frac{3+y}{2y}$, $p^{-1}(x) = \frac{3+y}{2y}$
    ii) 0.75

156 IGCSE Mathematics Revision Guide

c) i) $n(3) = 6(3) - 3 = 15 \rightarrow m(15) = 2(15^2) - 3 = 447$
    ii) $m(2) = 2(2^2) - 3 = 5 \rightarrow n(5) = 6(5) - 3 = 27$
d) i) $2(6x - 3)^2 - 3 = 2(36x^2 - 36x + 9) - 3 = 72x^2 - 72x + 15$
    ii) $6(2x^2 - 3) - 3 = 12x^2 - 18 - 3 = 12x^2 - 21$

Angle properties (p. 91)

Quick test
1. a) $a = 65°, b = 115°, c = 115°, d = 75°, e = 40°$
   b) $f = 40°, g = 30°$
2. a) $h = 20°, i = 110°, j = 20°$
   b) $k = 30°, l = 60°$
3. 24
4. 30

Geometrical terms and relationships (p. 97)

Quick test
1. $AC = 9.1\,\text{cm}, BC = 5.9\,\text{cm}$
2. 345°
3. Yes
4. The 4 small triangles are all congruent, but also triangles ABD and CBD are congruent, as are triangles ACD and CAB. Hence, there are six triangles.
5. They are not similar. The bases and the vertical sides both have a linear factor of 1.8, but the sloping sides have a linear factor of 1.9. To be similar all linear factors have to be the same.
6. 23.04 cm²
7. 43 mm
8. These triangles are congruent. By using Pythagoras, the missing side in the left-hand triangle is 12 cm and in the right-hand triangle 5 cm. Either SSS or RHS could be used as the reason.

Geometrical constructions (p. 99)

Quick test
1. Correct drawing
2. No. The sum of the two shorter sides must be greater than the longest side.
3.

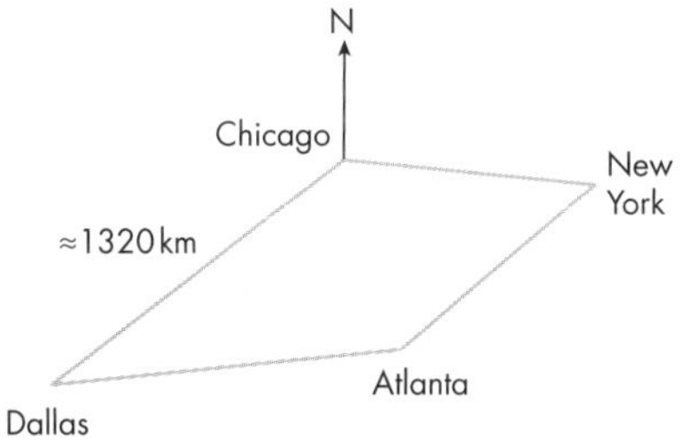

Trigonometry (p. 107)

Quick test
1. 10.82 cm
2. 14.46 cm
3. 0.8290, 0.5592, 1.4826
4. a) 68°   b) 34°
5. 4.09 cm, 25.8°, 6.88 cm
6. 4874
7. 64.9°
8. a) 105.0°   b) 170.0°
9. 46.4°, 25.83 cm²
10. 6.87 cm

Mensuration (p. 113)

Quick test
1. 3800 cm²
2. 68 cm²
3. 43 cm²
4. 6 cm
5. 63.6 cm²
6. 35 cm
7. 63 g (vol = 90 cm³)
8. 8.27 cm
9. Perimeter = $12\pi + 16$, area = 48π
10. 0.44 cm²
11. 6 cm
12. 158 cm³
13. 9.425 m²

Symmetry (p. 115)

Quick test
1. 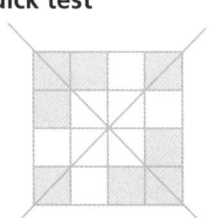
2. 7: H, I, N, O, S, X, Z
3. A rectangle has 2 lines of symmetry but a parallelogram has none.
4. 4, one from each vertex to the middle of the base.
5.

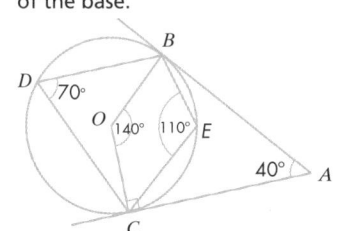

Vectors (p. 117)

Quick test
1. i) $m + n = \begin{pmatrix} 3 \\ -1 \end{pmatrix}$ ii) $3n = \begin{pmatrix} 3 \\ 6 \end{pmatrix}$
   iii) $m - n = \begin{pmatrix} 1 \\ -5 \end{pmatrix}$ iv) $n - m = \begin{pmatrix} -1 \\ 5 \end{pmatrix}$
   v) $2m + n = \begin{pmatrix} 5 \\ -4 \end{pmatrix}$

2. and 3.

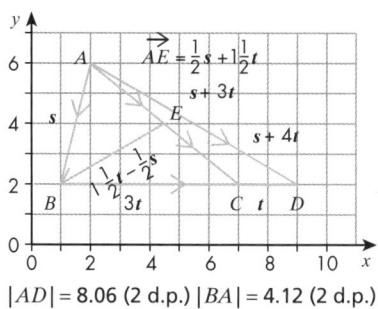

$|AD| = 8.06$ (2 d.p.) $|BA| = 4.12$ (2 d.p.)

Transformations (p. 122)

Quick test
1. a–e)

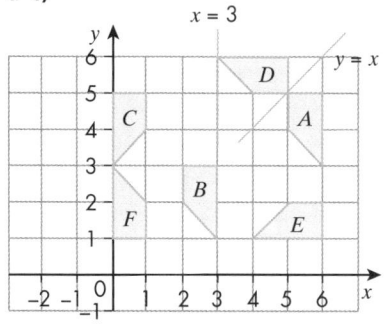

f) Reflection in $y = 3$
g) Reflection in $x + y = 6$
h) Rotation 90° clockwise about (3, 4)

2.

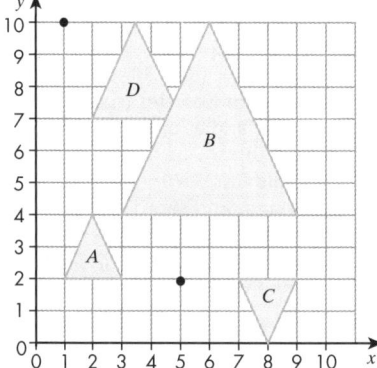

3.

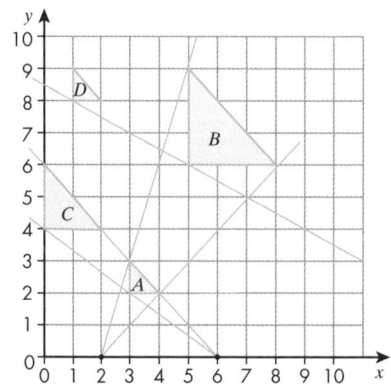

a) Scale factor 3, enlargement from (2, 0)
b) Scale factor 2, enlargement from (6, 0)
c) Scale factor 1/3, enlargement from (–1, 9)

Exam-style practice questions (p. 123–129)

1. a) i) $(10x - 7) + (2x + 7) = 180$, $12x = 180$, $x = 15$ so $2x + 7 = 37°$
   ii) $(3y - 25) + (3y - 5) = 180$, $6y - 30 = 180$, $6y = 210$, $y = 35$ so $3y - 5 = 100°$
   iii) $37 + 100 + z = 180$, $z = 43°$

   b) $(4x - 10) + (2x - 10) + (3x + 15) + (x + 15) = 360$, $10x + 10 + 360$, $10x = 350$, $x = 35$
   Natasha is right. It is a trapezium.
   A = 50°, B = 130°, C = 60°, D = 120°.
   If $ABCD$ were a parallelogram $A = C$ and $B = D$ but they do not.
   $A + B = 180°$ and $C + D = 180°$ making BC and AD parallel, and $ABCD$ a trapezium.

   c) $1880 = 180(n - 2) \to n = 12.4$, $1980 = 180(n - 2) \to n = 13$. Kim is right. The polygon has 13 sides.

   d) $\triangle ODC$ is isosceles so $\angle OCD = \frac{180 - 80}{2} = 50$. x is corresponding to $\angle OCD$ so $x = 50°$

   e) For x: $\angle BCD = 180 - 75 = 105$. $x = 75°$ — opposite angles of a cyclic quadrilateral add up to 180°
   For y: Angle $OCD = \frac{180 - 110}{2} = 35$, Angle $BCO = 180 - 75 - 35 = 70$. Angle $ABC = 110$ — interior angles add up to 180°, Angle $ADC = 70$ = opposite angle of a cyclic quadrilateral add up to 180°.
   $y = 70 - 35 = 35°$

   f) Angle $A = 180 - 70 - 70 = 40°$ (base angles of isosceles triangles are equal, angles of a triangle add up to 180)
   Angle $BAC = 70°$ (angle between tangent and chord = angle in alternate segment)
   Angle $BCO = 20°$ (radius meets tangent at 90°)
   Angle $ACB = \frac{180 - 70}{2} = 55°$ (base angle of isosceles triangles are equal)
   Angle $B = 55 - 20 = 35°$

2. a) $C = 30°$, $AC = 6.1$ cm
   b) 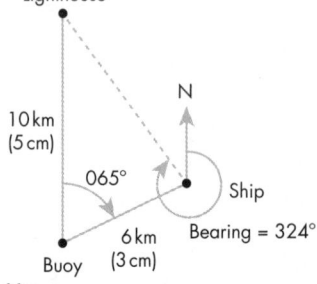 Lighthouse diagram: 10 km (5 cm), 065°, 6 km (3 cm), Buoy, Ship, N, Bearing = 324°
   c) Yes
   d) Angle $A = 70°$ and angle $D = 75°$. The angles in the two triangles do not match so they are not similar.
   e) Enlargement of length = $\sqrt{\frac{4500}{76.05}} = \frac{100}{13}$ Length of banknote = $100 \div \frac{100}{13} = 13$ cm
   f) Enlargement of volume = $\left(\sqrt{\frac{60}{18}}\right)^3$ = 6.09 Family carton = $180 \times 6.09 = 1096.2 \approx 1100$ ml
   g) 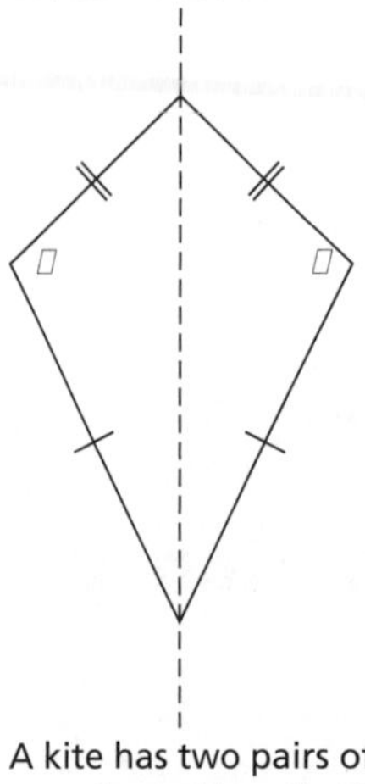 A kite has two pairs of sides that are of equal length. The line of symmetry is a shared side and the marked angles are equal. The two triangles are congruent: SSS or SAS.

3. a) 4.1 cm
   b) 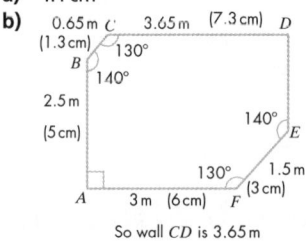 So wall CD is 3.65 m

4. a) $DB^2 = 14^2 - 8^2 = 132$, $DC = \sqrt{132 \times 5^2} = 10.3$ cm
   b) $DC = \sqrt{7^2 \times 10^2 \times 9^2} = 15.2$ cm
   c) Angle $ABD = \sin^{-1}\left(\frac{8}{14}\right) = 34.8°$
   d) Angle $C = \cos^{-1}\left(\frac{14}{16}\right) = 29.0°$
   e) Angle $BAC = \tan^{-1}\left(\frac{10}{7}\right) = 55.0°$
   f) Distance = $(52 - 1.7) \div \tan(72) = 17.3 \approx 17$ m
   g) 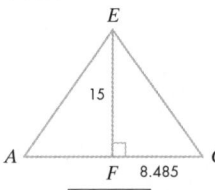 $AC = \sqrt{12^2 + 12^2} = 16.97$, $AF = 8.485$
   Angle $FEC = \tan^{-1}\left(\frac{8.485}{15}\right) = 29.495°$
   Angle $AEC = 29.495 \times 2 = 58.99 \approx 59.0°$
   h) $\frac{\sin(55)}{5.1} = \frac{\sin(B)}{6.2}$, $\sin(B) = \frac{\sin(59) \times 6.2}{5.1} = 0.9958$, $B = \sin^{-1}(0.9958) = 84.8°$
   i) $\frac{\sin(47)}{AB} = \frac{\sin(55)}{5.1}$, $AB = \frac{\sin(47) \times 5.1}{\sin(55)} = 4.55$ cm
   j) Area = $\frac{1}{2} \times 5.1 \times 6.2 \times \sin(47) = 11.56$ cm^2

5. a) 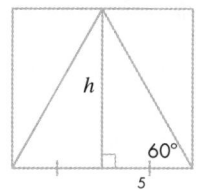 $h = \tan(60) \times 5 = 8.660$
   Area of rectangle = $10 \times 5.660 = 86.60$ cm^2
   b) Area of pond = $9 m^2$, Area = πr^2, $r = \sqrt{\frac{9}{\pi}}$, $r = 1.692$, $d = 3.385$, $c = \pi d = 10.63$ m
   c) 0.8 m = 80 cm, $\frac{80}{20} = 4$, $\frac{80}{8} = 10$, $\frac{80}{10} = 84 \times 10 \times 8 = 320$ toys
   d) i) Volume = $(\pi \times 1^2 \times 4) - (0.6 \times 0.6 \times 4) = 11.13$ cm^3
      ii) Mass = $11.13 \times 2.72 = 30.27 \approx 30$ g
   e) Volume = $\frac{60}{360} \times \pi \times 6^2 \times 4 = 75.398 \approx 75$ ml
   f) Volume = $\frac{360 - 70}{360} \times \pi \times 4^2 \times 2 = 80.98$ cm^2. Mass = $80.98 \times 3.5 = 283.43 \approx 280$ g
   g) i) Volume of cone = $\frac{1}{3} \times \pi \times 5^2 \times 12 = 100\pi$
      Volume of hemisphere = $\frac{2}{3} \times \pi \times 5^3 = \frac{250}{3}\pi$
      Total volume = $100\pi + \frac{250}{3}\pi = 575.95 \approx 576$ cm^3
   g) ii) Curved surface area = $\pi \times 5 \times 13 = 65\pi$
      Surface area of hemisphere = $2 \times \pi \times 52 = 50\pi$
      Total surface area = 115π

6. a) 8 lines of symmetry
   b) i) and ii) Angle $OAC =$ Angle $OBC = 90°$, $AC = BC$ so all 4 sides are equal length. It is a square. Angle $BOC =$ Angle $AOC = 45°$

7. a) 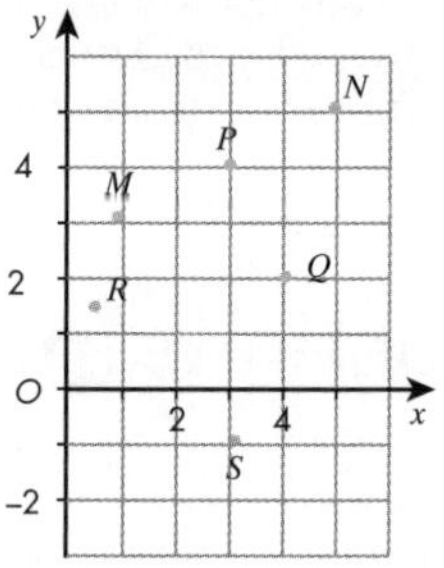 $\overrightarrow{MQ} = \begin{pmatrix} 3 \\ -1 \end{pmatrix}$
   $\overrightarrow{RN} = \begin{pmatrix} 4.5 \\ 3.5 \end{pmatrix}$
   $\overrightarrow{SP} = \begin{pmatrix} 0 \\ 5 \end{pmatrix}$
   $\overrightarrow{NS} = \begin{pmatrix} -2 \\ -6 \end{pmatrix}$
   $\overrightarrow{QR} = \begin{pmatrix} -3.5 \\ -0.5 \end{pmatrix}$
   $\overrightarrow{QP} = \begin{pmatrix} -1 \\ 2 \end{pmatrix}$

   b) i) $BD = BA + AD = -2x + 3y$
      ii) $AS = AB + \frac{1}{2}BD = 2x + \frac{1}{2}(-2x + 3y) = x + \frac{3}{2}y$
      iii) $\overrightarrow{AC} = 2x$, $\overrightarrow{AS} = 2x + 3y$
      $\overrightarrow{BC} = \overrightarrow{BA} + \overrightarrow{AC} = -2x + 2x + 3y = 3y$
      $\overrightarrow{BC}$ and $\overrightarrow{AD}$ are both $3y$, so they are parallel

c) $AC^2 = 1^2 + 7^2 = 1 + 49 = 50$
If right-angled triangle then Pythagoras' rule is satisfied.
$AC^2 = 50$
$BC^2 = 5 AB^2 = 45$
$50 = 45 + 5$
$AC^2 = BC^2 + AB^2$
Pythagoras' rule is satisfied, so triangle ABC is right angled. So $ABC = 90°$

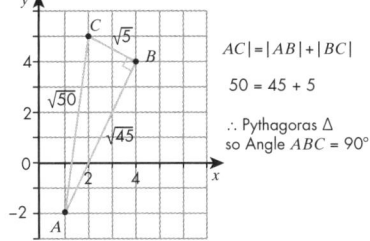

8. a)

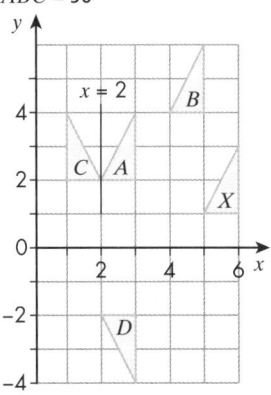

b)

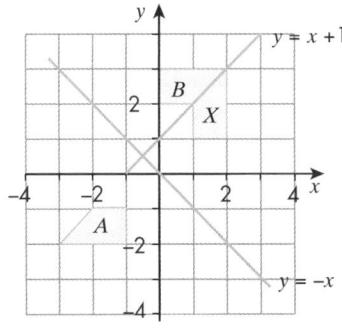

c)

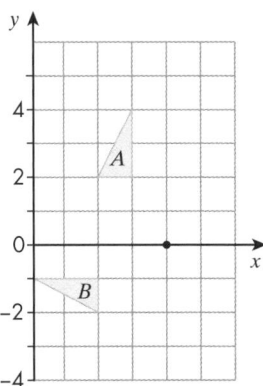

d)

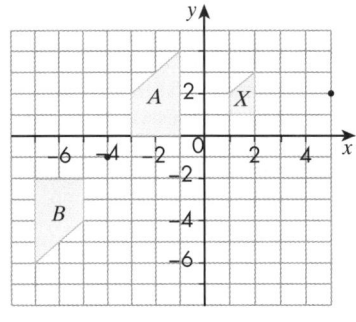

iii) Rotation of 180° about (−4, −1)

e)

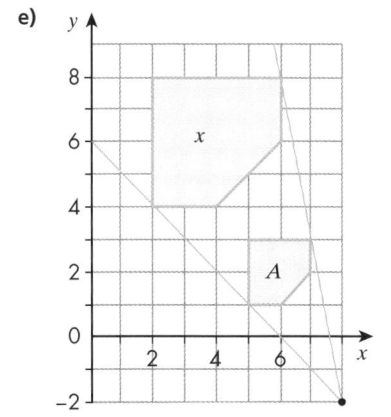

Statistical representation (p. 135)

Quick test

1. i)

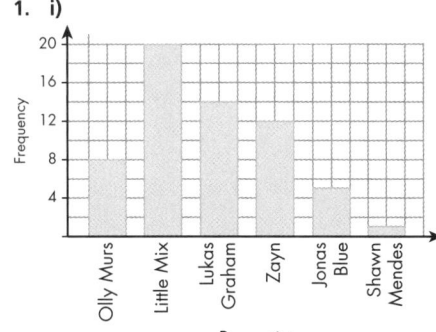

ii) Olly Murs $\quad \frac{8}{60} \times 360° = 48°$

Little Mix $\quad \frac{20}{60} \times 360° = 120°$

Lukas Graham $\quad \frac{14}{60} \times 360° = 84°$

Zayn $\quad \frac{12}{60} \times 360° = 72°$

Jonas Blue $\quad \frac{5}{60} \times 360° = 30°$

Shawn Mendes $\quad \frac{1}{60} \times 360° = 6°$

2.

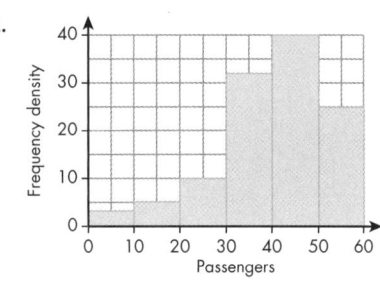

3.
```
4 | 1 2 7
5 | 3 4 6 8
6 | 3 8 8 9
7 | 1 3 4 6
8 | 1 2 4 6 8 9
9 | 0 1 4 5    key 4|1 means 41 passengers
```
Median = 73 passengers
Mode = 68 passengers

4. $0 < s \leq 10 \quad 19 \div 10 = 1.9$
$10 < s \leq 15 \quad 24 \div 5 = 4.8$
$15 < s \leq 20 \quad 32 \div 5 = 6.4$
$20 < s \leq 25 \quad 29 \div 5 = 5.8$
$25 < s \leq 35 \quad 26 \div 10 = 2.6$
$35 < s \leq 60 \quad 20 \div 25 = 0.8$

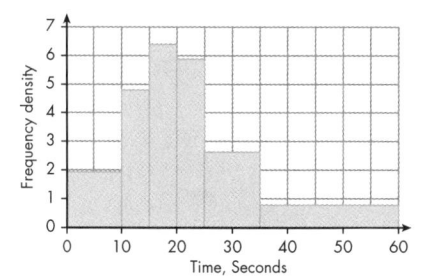

Statistical measures (p. 142)

Quick test

1. a) mode = 36, median = 29.5, mean = 31.5, range = 28
   b) mode = 6, median = 6, mean = 5.63, range = 10

2. 233.25 cm²

3. Median = 235, LQ = 185, UQ = 285, IQR = 100

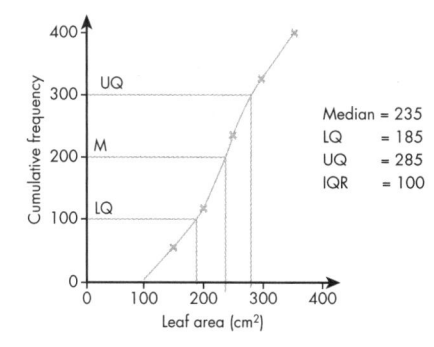

4.

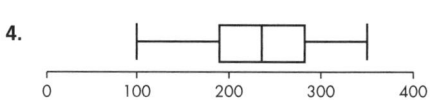

Probability (p. 147)

Quick test

1. a) $\frac{17}{50}$ b) $\frac{46}{50} = \frac{23}{25}$
   c) $\frac{19}{50}$ d) 0

2. 0.18

3. a) P(white) = 0.6
   b) 10 trials, P(white) = 0.4;
   50 trials;
   P(white) = 0.62;
   100 trials;
   P(white) = 0.48;
   250 trials;

P(white) = 0.52;
500 trials;
P(white) = 0.59

c) 500 trials gives the best approximation to the theoretical probability. The more trials that are completed, the closer the experimental probability.

4. a) $\frac{2}{12} = \frac{1}{6}$ b) $\frac{9}{12} = \frac{3}{4}$ c) $\frac{9}{12} = \frac{3}{4}$

		Spinner A		
		R	R	B
Spinner B	R	R, R	R, R	B, R
	B	R, B	R, B	B, B
	B	R, B	R, B	B, B
	G	R, G	R, G	B, G

5. a) 0.64 b) 0.32

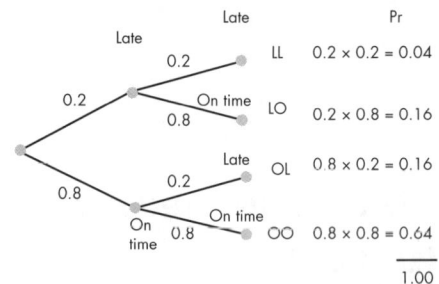

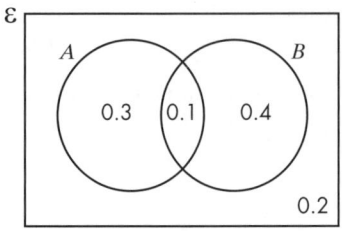

Tree diagram:
- 0.2 Late → 0.2 Late: LL 0.2 × 0.2 = 0.04
- 0.2 Late → 0.8 On time: LO 0.2 × 0.8 = 0.16
- 0.8 On time → 0.2 Late: OL 0.8 × 0.2 = 0.16
- 0.8 On time → 0.8 On time: OO 0.8 × 0.8 = 0.64
- Total = 1.00

6. a) Venn diagram: A = 0.3, A∩B = 0.1, B = 0.4, outside = 0.2

b) i) 0.5 ii) 0.8 iii) 0.1

7. a) 0.21 b) 0.52 c) 0.34

Exam-style practice questions (p. 148–151)

1. a)

Make	Tally	Frequency
VW	|||| |||	8
Ford	|||| ||||	13
Vauxhall	|||| |||	8
Suzuki	||||		6
Renault	|||| |||	8
Peugeot	|||| |||| ||	12
Mercedes	||||	5
		60

Mode = Ford

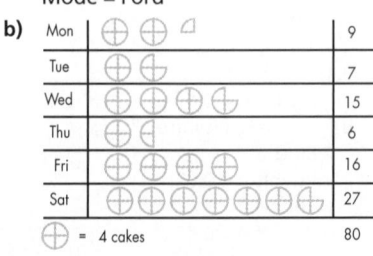

b)
Mon	⊕⊕◁	9
Tue	⊕⊖	7
Wed	⊕⊕⊕⊖	15
Thu	⊕⊖	6
Fri	⊕⊕⊕⊕	16
Sat	⊕⊕⊕⊕⊕⊕⊖	27

⊕ = 4 cakes   80

80 cakes were sold in the week.

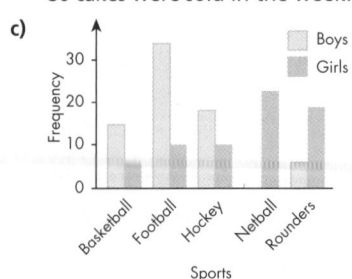

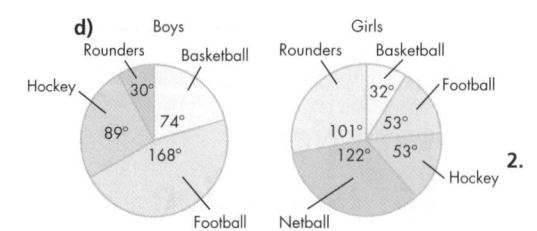

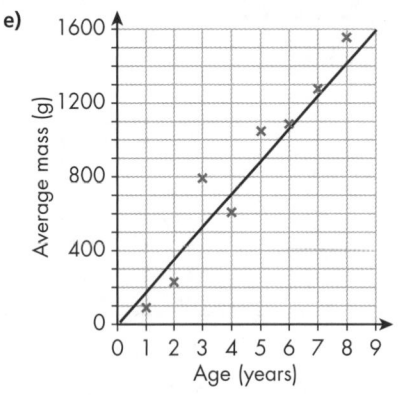

c) Bar chart: Boys and Girls frequencies for Basketball, Football, Hockey, Netball, Rounders.

d) Boys pie chart: Rounders 30°, Basketball 74°, Football 168°, Hockey 89°.
Girls pie chart: Rounders 32°, Basketball 53°, Football 53°, Netball 122°, Hockey 101°.

e) Scatter graph of Average mass (g) vs Age (years).

The average mass for a 3-year-old koi (795 g) seems to be in error.
9 years ≈ 1600 g

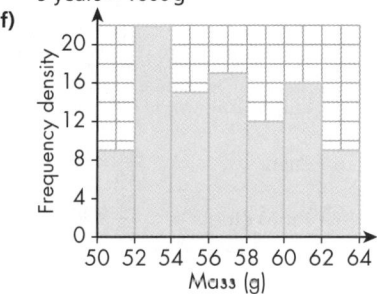

f) Histogram of Frequency density vs Mass (g) from 50 to 64.

g)
```
6 | 1 2 2 4 6 7
7 | 0 1 3 4 4 5 8 8 9
8 | 1 2 2 3 5 6 6 6 7 8
9 | 2 2 3 6              key 6|1 means 61 kg
```
Mode = 86 kg
Median = 79 kg

h) The class intervals are unequal so the y axis will represent 'Frequency density'.

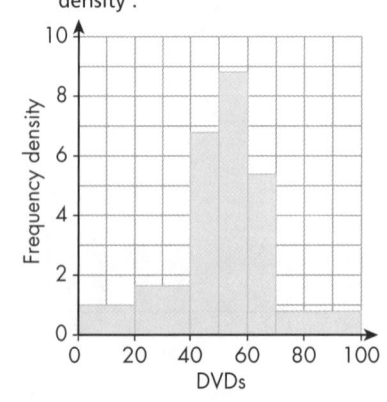

Histogram: Frequency density vs DVDs (0 to 100).

Frequency density = $\frac{frequency}{class\ interval}$, so for example,

$0 < d \leq 20$ has a $fd = \frac{16}{20} = 0.8$

25 DVDs: 16 students are in the $0 < d \leq 20$ class interval and the 25th DVD occurs partway $\left(\frac{5}{20}\right)$ through the class interval $20 < d \leq 40$. Since there are 34 students in this class interval the student estimate for the 25th DVD in the $20 < d \leq 40$ interval is: $\frac{5}{20} \times 34 = 8.5$

The number of students with < 25 DVDs can be estimated as: $16 + 8.5 = 24.5$, so 24 people.

2. a) Mode = 21, median = 20.5, mean = 20, range = 12
b) Mode = 12, median = 13, mean = 12.7, range = 5
c) Mean = $[(25 \times 3) + (75 \times 6) + (125 \times 11) + (175 \times 7) + (225 \times 2) + (275 \times 1)] \div 30 = 3850 \div 30 = 128.33 \approx 128$ runs
Median: $100 \leq s < 150$
Modal group: $100 \leq s < 150$

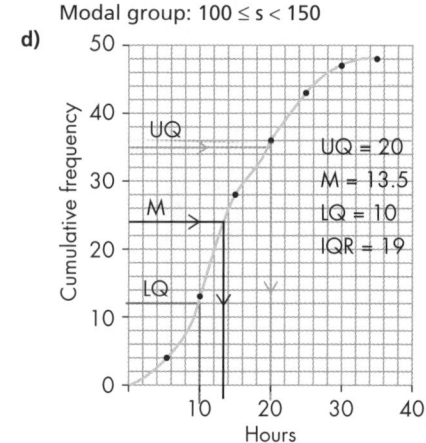

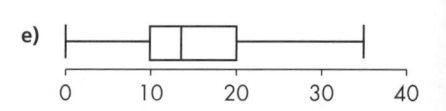

d) Cumulative frequency curve: UQ = 20, M = 13.5, LQ = 10, IQR = 19

e) Box plot from approximately 0 to 35, median ~13.

3. a) i) $\frac{1}{10}$ ii) $\frac{1}{5}$ iii) $\frac{1}{2}$
iv) $\frac{1}{10}$ v) $\frac{1}{100}$ vi) $\frac{1}{25}$

b) i) $\frac{7}{11}$ ii) $\frac{4}{11}$ iii) $\frac{9}{11}$
iv) $\frac{4}{11}$

c) Each side of a 10-sided spinner has a probability of 0.1, meaning each side should occur 25 times out of 250. Each side of an 8-sided spinner has a probability of 0.125, meaning each side should also occur 25 times out of 200. Tanya's spinner seems to be the one with least bias because each number occurs within ± 5 of this. Mikhail's spinner seems biased, with the number 2 occuring too few times and the number 6 far too frequently.

d) i)

	Y	Y	Y	B	B
Y	YY	YY	YY	BY	BY
Y	YY	YY	YY	BY	BY
Y	YY	YY	YY	BY	BY
B	YB	YB	YB	BB	BB
B	YB	YB	YB	BB	BB

ii) P(2 yellow counters) = $\frac{9}{25}$,

P(2 blue counters) = $\frac{4}{25}$,

P(1 yellow and 1 blue counter) = $\frac{12}{25}$,

P(blue followed by a yellow counter) = $\frac{6}{25}$

e)

Throw 1 — Throw 2
- 75% Hit → 90% Hit: HH 67.5%
- 75% Hit → 10% Miss: HM 7.5%
- 25% Miss → 40% Hit: MH 10%
- 25% Miss → 60% Miss: MM 15%

Total 100%

i) 67.5%
ii) 17.5%
iii) 450

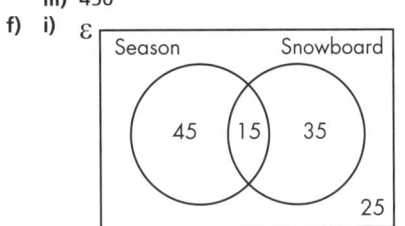

f) i)

ε — Season and Snowboard Venn diagram: Season only 45, intersection 15, Snowboard only 35, outside 25

ii) $\frac{3}{8} \left(\frac{45}{120} \right)$

g) i) $\left(\frac{14}{20} \times \frac{6}{19} \right) + \left(\frac{6}{20} \times \frac{14}{19} \right)$

$= \frac{168}{380} \left(\frac{42}{95} \right)$

ii) $1 - P(\text{2 skiers}) = 1 - \left(\frac{14}{20} \times \frac{13}{19} \right)$

$= \frac{198}{380} \left(\frac{99}{190} \right)$

Glossary

acceleration – The rate at which the velocity of a moving object increases.

acute angle – an angle that measures less than 90°.

acute-angled triangle – A triangle in which all the angles are acute.

adjacent side – The side that is between a given angle and the right angle in a right-angled triangle.

algebraic fraction – A fraction that includes algebraic terms.

allied angles – Interior angles that lie on the same side of a transversal that cuts a pair of parallel lines; they add up to 180°.

alternate angles – Angles that lie on either side of a transversal that cuts a pair of parallel lines; the transversal forms two pairs of alternate angles and the angles in each pair are equal.

angle bisector – A line or line segment that divides an angle into two equal parts.

angle of depression – The angle between the horizontal line of sight of an observer and the direct line to an object that is viewed from above.

angle of elevation – The angle between the horizontal line of sight of an observer and the direct line to an object that is viewed from below.

angles around a point – The angles formed at a point where two or more lines meet; their sum is 360°.

angles on a straight line – The angles formed at a point where one or more inclined (sloping) lines meet on one side of a straight line; their sum is 180°.

approximate – A value that is close but not exactly equal to another value, which can be used to give an idea of the size of the value; for example, a journey taking 58 minutes may be described as 'taking approximately an hour'; the ≈ sign means 'is approximately equal to'.

arc – Part of the circumference of a circle.

area rule – The rule for the area of triangle, $A = \frac{1}{2}ab\sin C$, where a and b are two sides of the triangle and C is the included angle.

area scale factor – The ratio of the area of one shape to the area of another that is mathematically similar to it.

asymptote – A line that a curve approaches but never quite meets.

average speed – The result of dividing the total distance travelled by the total time taken for a journey.

bar chart – A chart using bars or columns of equal width to represent frequency.

bias – The property of a sample being unrepresentative of the population; for example, a dice may be weighted so that it gives a score of 5 more frequently than any other score.

BIDMAS/BODMAS – An acronym to help you remember the order of arithmetic operations: Brackets, Index (/Orders), Divide, Multiply, Add, Subtract

bisect – Cut exactly in half.

capacity – The amount of space in a container; the amount of liquid it can hold.

centre of enlargement – The point, inside or outside the object, on which an enlargement is centred; the point from which the enlargement of an object is measured.

centre of rotation – The point about which an object or shape is rotated.

circumference – The perimeter of a circle; every point on the circumference is the same distance from the centre, and this distance is the radius.

class interval – a group of values in a set of grouped data.

coefficient – A number or constant term written in front of a variable in an algebraic term; for example, in $8x$, 8 is the coefficient of x.

combined event – Two or more events that occur together or sequentially.

common factor – A factor that divides exactly into two or more numbers; for example, 2 is a common factor of 6, 8 and 10.

completing the square – an alternative method for solving a quadratic equation which changes $x^2 + px + q$ into the form $(x + a)^2 + b$. The conversion is: $\left(x + \frac{p}{2}\right)^2 - \left(\frac{p}{2}\right)^2 + q$.

composite – A function that is made from two or more separate functions.

compound interest – Interest that is paid on the amount in the account and then added to the account total; after the first year, interest is paid on interest earned in the previous years.

congruent – Exactly the same shape and size.

constant of proportionality – If two variables x and y are in direct proportion, you can write an equation, $y = kx$; if they are in inverse proportion, you can write $xy = k$. In either case, k is the constant of proportionality.

continuous data – Data, such as mass, length or height, that can take any value; continuous data has no precise fixed value.

conversion graph – A graph that can be used to convert from one unit to another, constructed by drawing a line through two or more points where the equivalence is known; sometimes, but not always, a conversion graph passes through the origin.

corresponding angles – Angles that lie on the same side of a pair of parallel lines cut by a transversal; the transversal forms four pairs of corresponding angles, and the angles in each pair are equal.

cosine – A trigonometric ratio related to an angle in a right-angled triangle, calculated as $\frac{\text{adjacent}}{\text{hypotenuse}}$.

cosine rule – A rule relating the cosine of one angle in a triangle to the lengths of all three sides. $a^2 = b^2 + c^2 - 2bc \cos A$ or $\cos A = \frac{b^2+c^2-a^2}{2bc}$.

cross-section – A cut across a 3D shape, or the shape of the face that is exposed when a 3D shape is cut. For a prism, a cut across the shape, perpendicular to its length.

cube number – The result of multiplying a number by itself and then by the original number again. Written as x^3.

cube root – The cube root of a number is another number which, when cubed, will equal the first number. Written as $\sqrt[3]{\ }$. For example, 2^3 is 8, and the cube root of 8 is 2.

cumulative frequency – The total frequency of all values up to the end of each class interval; a running total.

cyclic quadrilateral – A quadrilateral with vertices that lie on the circumference of a circle; the sum of opposite angles is 180°

cylinder – A prism with a circular cross-section.

decimal fraction – a fraction written using the normal decimal place-value system with 10, 100, 1000 and so on as the denominators.

decimal place – The position, after the decimal point, of a digit in a decimal number; for example, in 0.025, 5 is in the third decimal place. Also, the number of digits to the right of the decimal point in a decimal number; for example, 3.142 is a number given correct to three decimal places (3 d.p.).

density – The mass of a substance divided by its volume.

difference of two squares – An expression of the form $x^2 - y^2$; the terms are squares and there is a minus sign between them.

direct proportion – A relationship in which one variable increases or decreases at the same rate as another; in the formula $y = 12x$, x and y are in direct proportion.

direct variation – Another name for direct proportion.

discrete data – Data that can only take certain values, such as a number of children; discrete data can only take fixed, separate values.

distance–time graph – A graph that represents a journey, based on the distance travelled and the time taken.

element – Any single member of a set.

eliminate – Given a pair of simultaneous equations with two variables, you can manipulate one or both equations to remove or eliminate one of the variables by a process of substitution, addition or subtraction.

enlargement – A transformation in which the object is enlarged to form an image.

equation – A relationship in which two expressions are separated by an equals sign with one or more variables. An equation can be solved to find one or more answers, but it may not be true for all values of x.

equilateral triangle – A triangle in which all the sides are equal and all the angles are 60°

equivalent fraction – A fraction with the same value as another, for example, $\frac{1}{2} = \frac{2}{4}$, $\frac{2}{3} = \frac{6}{9}$

event – Something that happens in a probability problem, such as tossing a coin or predicting the weather.

exhaustive – All possible outcomes of an event; the sum of the probabilities of exhaustive outcomes equals 1.

expand – Multiply out (terms with brackets).

experimental probability – An estimate for the theoretical probability based on the number of experiments (or trials) carried out.

exponential growth/decay – A rate of change is exponential. Use the compound interest formula.

expression – A collection of numbers, letters, symbols and operators representing a number or amount; for example, $x^2 - 3x + 4$

exterior angle – The angle formed outside a 2D shape, when a side is extended beyond the vertex.

factor – A number that will divide exactly into another.

factorisation – The arrangement of a given number or expression into a product of its factors. (verb: to factorise)

formula – A mathematical rule, using numbers and letters, that shows a relationship between variables; for example, the conversion formula from temperatures in Fahrenheit to temperatures in Celsius is: $C = \frac{5}{9}(F - 32)$.

fraction – A part of a whole that has been divided into equal parts; a fraction describes how many parts you are talking about.

frequency – The number of times that something is counted or observed.

frequency density – The frequency of a class interval divided by the width of the class interval, used when drawing histograms.

frustum – A 3D shape produced by removing the top from a pyramid or cone, by means of a cut parallel to the base.

function – An algebraic expression in which there is only one variable, often x. $f(x)$ is the symbol for a function with a single variable, x.

gradient – The amount of slope of a line between two or more points, calculated as the vertical difference between the coordinates divided by the horizontal difference.

grouped data – Data arranged into smaller, non-overlapping sets, groups or classes, that can be treated as separate ranges or values, for example, 1–10, 11–20, 21–30, 31–40, 41–50; in this example there are equal class intervals.

highest common factor (HCF) – The largest number that is a factor common of two or more other numbers.

histogram – A diagram, similar to a bar chart, in which the area of each bar is proportional to the frequency of its class interval.

hypotenuse – The longest side in a right-angled triangle, always opposite the right angle.

image – The result of a reflection or other transformation of an object.

improper fraction – A fraction in which the denominator is greater than the numerator, also called a vulgar fraction.

index – The power to which a base number is raised; in 3^4, 4 is the index and 3 is the base number.

index notation – Expressing a number in terms of one or more of its factors, each expressed as a power.

inequality – A statement that one expression is greater or less than another, written with the symbol > (greater than) or < (less than) instead of = (equals).

integer – A whole number, positive or negative and including zero.

intercept – The point where a line cuts or crosses an axis.

interior angle – The inside angle between two adjacent sides of a 2D shape, at a vertex.

interquartile range – A measure of dispersion calculated as the upper quartile minus the lower quartile, often abbreviated to IQR.

intersection – The 'overlap', the set of elements that occur in two or more sets.

inverse – Going the other way.

inverse function – Reverse or opposite; inverse operations cancel each other out or reverse the effect of each other.

inverse operation – An operation that reverses the effect of another operation; for example, addition is the inverse of subtraction, division is the inverse of multiplication.

inverse proportion – A relationship between two variables in which as one value increases, the other decreases; in the formula $xy = 12$, x and y are in inverse proportion.

inverse variation – Another name for *inverse proportion*.

irrational number – Numbers that are never-ending **and** non-repeating, for example, $\sqrt{2}$ or π. They cannot be written as a fraction.

isosceles triangle – A triangle in which two sides are equal and the angles opposite the equal sides are also equal.

like terms – Terms in which the variables are identical, but the coefficients are different; for example, $2ax$ and $5ax$ are like terms but $5xy$ and $7y$ are not. Like terms can be combined by adding their numerical coefficients so $2ax + 5ax = 7ax$.

limits of accuracy – No measurement is entirely accurate. The accuracy depends on the tool used to measure it. The value of every measurement will be rounded to within certain limits. For example you can probably measure with a ruler to the nearest half-centimetre. Any measurement you take could be inaccurate by up to half a centimetre. This is your limit of accuracy. (See also *lower bound and upper bound*.)

line bisector – A line that divides another line exactly in half.

line of best fit – A straight line drawn on a scatter diagram where there is correlation, so that there are roughly equal numbers of points above and below it; the line shows the trend of the data.

linear – An expression (such as $5x + 2$) for which there is a term with an index of 1 and possibly a constant term.

linear scale factor – The factor of increase between the lengths of two similar shapes.

linear sequence – A sequence or pattern of numbers in which the difference between consecutive terms is always the same.

loci – The plural of locus.

locus – The path of a point that moves obeying given conditions.

lower bound – The lower limit of a measurement. (See also *limit of accuracy*.)

lower quartile – The lowest value of the three quartiles, often abbreviated to Q1.

lowest common multiple (LCM) – The lowest number that is a multiple of two or more numbers; for example, 12 is the lowest common multiple of 2, 3, 4 and 6.

magnitude – The size of a quantity.

mass – The amount of matter in an object.

mean – Generally called the average, it is a calculation of the sum of all the data values divided by the number of data values.

median – The middle value of an ordered data set.

minimum point – A point on a graph where the gradient is zero, and which is lower than the points either side of it.

mirror line – Another name for a line of symmetry.

mixed number – A whole number and a proper fraction, for example, $2\frac{1}{2}$

modal group – In grouped data, the class with the highest frequency.

multiple – If a number divides into another number, the second is a multiple of the first.

multiplier – A number that is used to find the result of increasing or decreasing an amount by a percentage.

natural number – A positive integer.

negative correlation – A relationship between two sets of data, in which the values of one variable increase as the values of the other variable decrease.

net – A 2D shape or pattern that can be folded to make a 3D shape.

no correlation – No relationship between two sets of data.

number sequence – A sequence of numbers that follow a rule.

nth term – An expression in terms of n; it allows you to find any term in a sequence, without having to use a term-to-term rule.

object – The original or starting shape, line or point before it is transformed to give an image.

obtuse angle – an angle that measures between 90° and 180°.

opposite side – The side that is opposite a given angle, in a right-angled triangle.

order of rotational symmetry – The number of times a 2D shape looks the same as it did originally when it is rotated through 360° about a central point. If a shape has no rotational symmetry, its order of rotational symmetry is 1, because every shape looks the same at the end of a 360° rotation as it did originally.

outcome – A possible result of an event in a probability experiment, such as the different scores when throwing a dice.

parabola – The shape of a quadratic curve.

parallel – Lines that stay exactly the same distance apart, however far they are extended.

percentage – A measure out of 100; a fraction with a denominator of 100, for example, $\frac{47}{100} = 47\%$

percentage change – A change to a quantity, calculated as a percentage of the original quantity.

percentage loss – The loss in a financial transaction, calculated as the difference between the buying price and the selling price, given as a percentage of the original price.

perpendicular – Meeting at 90°.

perpendicular bisector – A line that divides a given line exactly in half, passing through its midpoint at right angles to it.

pi (π) – The result of dividing the circumference of a circle by its diameter, represented by the Greek letter pi (π).

pictogram – A chart that uses symbols to represent frequency.

pie chart – A chart that is a circle split into sectors, each representing frequency as a proportion.

polygon – A closed 2D shape with straight sides.

positive correlation – A relationship between two sets of data, in which the values of one variable increase as the values of the other variable increase.

prime factor – Factors of a number that are prime numbers.

prime number – A number that has exactly two factors: 1 and itself. Note that 1 is not a prime number.

prism – A 3D shape that has the same cross-section wherever it is cut perpendicular to its length.

probability – The chance or likelihood of an event happening; may be written as a fraction, decimal or percentage, and is always between 0 and 1

pyramid – A 3D shape with a base and sides rising to form a single point.

Pythagoras' theorem – The rule that, in any right-angled triangle, the square of the hypotenuse is equal to the sum of the squares of the other two sides.

quadratic expression – An expression of the form $ax^2 + bx + c$ in which the highest power of any variable is 2, such as $2x^2 + 4$

quadratic formula – A formula used to solve quadratic equations of the form $ax^2 + bx + c = 0$,
$x = \frac{-b \pm \sqrt{b^2 - 4ac}}{2a}$

quadrilateral – A four-sided 2D shape.

quartile – One of three points that divides a set of data, in numerical order, into four equal parts.

random – Chosen by chance; every item has an equal chance of being chosen.

range – The difference between the largest and smallest values in a data set.

ratio – The ratio of A to B is a number found by dividing A by B. It is written as A : B. For example, the ratio of 1 m to 1 cm is written as 1 m : 1 cm = 100 : 1. Notice that the two quantities must both be in the same units if they are to be compared in this way.

rational number – A number that can be written as a fraction exactly; an integer, fraction or recurring decimal.

real numbers – The set of numbers made up of rational and irrational numbers.

rearrange – Put into a different order, to simplify.

reciprocal – The result of dividing a number into 1, so 1 divided by the number is its reciprocal.

recurring decimal – A decimal number in which a digit or pattern of digits repeats for ever.

reflection – The image formed when a 2D shape is reflected in a mirror line or line of symmetry; the process of reflecting an object.

reflex angle – an angle that measures more than 180°

region – An area bound by inequalities.

relative frequency – An estimate of theoretical probability.

right angle – an angle measuring 90°.

right-angled triangle – A triangle in which one angle is 90°.

roots – The points on a graph where it crosses the x-axis.

rotate – To turn about a central point, called the centre of rotation.

rotational symmetry – A type of symmetry in which a 2D shape may be turned through an angle so that it looks the same as it did originally in two or more positions.

rounding – Expressing a number to a required degree of accuracy, such as decimal places or significant figures.

sample space diagram – A diagram that shows all the outcomes of an experiment.

scale drawing – A drawing that represents something much larger or much smaller, in which the lengths on the image are in direct proportion to the lengths on the object.

scale factor – The ratio of the distance on the image to the distance it represents on the object; the number that tells you how much a shape is to be enlarged.

scalene triangle – A triangle in which all sides are different lengths and all angles are different sizes.

scatter diagram/scattergraph – A graphical representation showing whether there is a relationship between two sets of data.

sector – A region of a circle, like a slice of a pie, enclosed by an arc and two radii.

segment – the shape that is made between a chord of a circle and one of the arcs that join the ends of the chord.

sequence – A pattern of numbers that are related by a rule.

set – A collection of objects or elements.

significant figure – In the number 12 068, 1 is the first and most significant figure and 8 is the fifth and least significant figure. In 0.246 the first and most significant figure is 2. Zeros at the beginning or end of a number are not significant figures.

similar – Two shapes are similar if one is an enlargement of the other; angles in the same position in both shapes are equal to each other.

similar triangles – Two or more triangles where one is an enlargement of the other.

simple interest – Money that a borrower pays a lender, for allowing them to borrow money. Based on the formula: principal × rate of interest × length of time in years ÷ 100

simplify – To make an equation or expression easier to work with or understand by combining like terms or cancelling; for example: $4a - 2a + 5b + 2b = 2a + 7b$, $\frac{12}{18} = \frac{2}{3}$, $5:10 = 1:2$.

simultaneous equations – Two equations that are both true for the same set of values for their variables.

sine – A trigonometric ratio related to an angle in a right-angled triangle, calculated as $\frac{opposite}{hypotenuse}$.

sine rule – A rule using sines of angles in any triangle showing that the ratio of the sine of an angle to the length of the side opposite it is always the same for any given triangle.
$\frac{a}{\sin A} = \frac{b}{\sin B} = \frac{c}{\sin C}$

slant height – The length of the sloping side of a cone.

sphere – A 3D shape which is the locus of a point that moves a fixed distance from a given point, the centre; a 3D shape that has a circular cross-section whenever it is cut through its centre.

square number – A number that may be obtained by multiplying another number by itself, written as x^2.

square root – Another number that, when squared, gives the first number; written as $\sqrt{}$.

standard form – A way of writing a number as $A \times 10^n$, where $1 \leq A < 10$ and n is a positive or negative integer.

subject – The variable on the left-hand side of the equals (=) sign in a formula or equation.

substitute – Replace a variable in an expression with a number and evaluate it; for example, if you substitute 4 for t in $3t + 5$ the answer is 17 because $3 \times 4 + 5 = 17$.

subtend – The joining of the lines from two points giving an angle.

surd – An irrational number found by taking a root of a number such as the square root of 2 ($\sqrt{2}$) or the cube root of 5 ($\sqrt[3]{5}$).

surface area – The total area of all of the surfaces of a 3D shape.

tangent – 1 A straight line that touches a circle just once.

2 – trigonometric ratio related to an angle in a right-angled triangle, calculated as $\frac{opposite}{adjacent}$.

term – 1 A part of an expression, equation or formula. Terms are separated by + and − signs.

2 – number in a sequence or pattern.

terminating decimal – A terminating decimal can be written down exactly. $\frac{33}{100}$ can be written as 0.33, but $\frac{1}{3}$ is 0.3333… with the digit 3 recurring forever.

theoretical probability – The exact or true probability of an event happening.

three-figure bearing – The angle measured from north clockwise, generally given as a three-digit figure.

transform – Reflect, rotate, enlarge or translate.

transformation – A change to a geometric 2D shape, such as a translation, rotation, reflection or enlargement.

translation – A movement along, up or diagonally on a coordinate grid.

travel graph – Another name for a distance–time graph.

tree diagram – A diagram that is used to calculate the probability of combined events happening. All the probabilities of each single event are written on the branches of the diagram.

trial – A single experiment in a probability experiment.

turning point – Any point on a graph where the gradient is zero; could be a maximum or a minimum.

union – The set of all the elements that occur in one or more sets.

unitary method – A method of finding best value by finding the price per unit, or the quantity per dollar or cent.

universal set – The set that contains all possible elements, usually represented by the symbol ε.

upper bound – The higher limit of a measurement. (See also *limit of accuracy*.)

upper quartile – The highest value of the three quartiles, often abbreviated to Q3.

variable – A letter that stands for a quantity that can take various values.

vector – A quantity such as velocity that has magnitude and acts in a specific direction.

velocity–time graph – A graph in which distance travelled is plotted against time taken.

Venn diagram – A diagram that shows the relationships between different sets.

vertex – The point at which two lines meet, in a 2D or 3D shape.

vertical height – The height of the top vertex of a 3D shape, measured perpendicular to the base.

vertically opposite angles – The angles on the opposite side of the point of intersection when two straight lines cross, forming four angles. The opposite angles are equal.

vertices – The plural of vertex.

volume scale factor – The factor of increase between the volumes of two similar shapes.

vulgar fraction – another term meaning exactly the same as *fraction*.

$y = mx + c$ – The general equation of a straight line in which m is the gradient of the line and c is the intercept on the y-axis.